America Project Making of, Fort Wayne Pittsburgh

A Compilation of the Laws, Deeds, Mortgages, Leases, And Other Instruments Affecting the Pittsburgh, Fort Wayne and Chicago Railway Company

America Project Making of, Fort Wayne Pittsburgh

A Compilation of the Laws, Deeds, Mortgages, Leases, And Other Instruments Affecting the Pittsburgh, Fort Wayne and Chicago Railway Company

ISBN/EAN: 9783744724722

Printed in Europe, USA, Canada, Australia, Japan

Cover: Foto ©berggeist007 / pixelio.de

More available books at **www.hansebooks.com**

A COMPILATION

OF THE

LAWS, DEEDS, MORTGAGES, LEASES,

AND OTHER INSTRUMENTS,

AND

MINUTES OF PROCEEDINGS,

AFFECTING

The Pittsburgh, Fort Wayne and Chicago Railway Company,

TOGETHER WITH

A PREFATORY STATEMENT

BY THE CHAIRMAN.

COMPILED BY DIRECTION OF THE PURCHASING COMMITTEE.

———

NEW YORK:
JOHN POLHEMUS, PRINTER.
[102 Nassau Street]
———
1875.

CONTENTS.

I.

II.

PAGE

III.

IV.

V.

PREFATORY STATEMENT.

The Purchasing Committee of the Pittsburgh, Fort Wayne and Chicago Railroad, having watched over the interests of all concerned in that undertaking for upwards of fourteen years, either as members of the Committee named, or of the Board of Directors of the Railway Company, are satisfied that their trust in the former capacity may now be safely closed. Preparatory to doing this, they have caused the present volume to be prepared, for the purpose of furnishing to the stockholders the means of referring to the text of the various securities which represent the property, and of informing themselves accurately in respect to the corporate rights and franchises under which the railroad is operated, and the individual rights of all interested therein.

Shortly after the organization of the present Company, the Committee caused a pamphlet to be made up, containing the principal documents in existence at that time, and the Board of Directors, also, have, from time to time, had particular papers printed and circulated, in pamphlet form, for the use of the stock and bondholders. But these pamphlets do not contain, singly or together, all the important documents affecting this Company; and they have become so scarce as practically to be inaccessible to the bondholders and stockholders at large. The Committee, on this account, determined that a new compilation was advisable. They considered also, that a book recording the documents and transactions which show the history and progress of this great road would be referred to with interest, by those who have invested in its capital or securities, or who may desire to do so hereafter,—many of whom may not be aware of the extraordinary transitions through which the enterprise has passed, or of the 'policy by which it has been lifted to its present prosperity.

When the Purchasing Committee entered actively upon their duties, the First Mortgage Bonds, then secured on sections of the road, and which are now represented by the First

Mortgage Bonds of the Pittsburgh, Fort Wayne and Chicago Railway Company, were selling at about forty-five cents on the dollar, although bearing about twenty-seven per cent. of accumulated interest. The bonds issued in substitution for these dishonored securities are now sought after, by sagacious and discriminating investors, at a premium ranging from twelve to fourteen per cent. So that they may be considered equal to any investment, whatever, public or private.

When the Purchasing Committee were appointed, nine different classes of bonds (including the Allegheny River Bridge Bonds), secured upon the railroad or some section thereof; two classes of bonds secured upon real estate held for depot purposes, &c., and numerous coupon bonds, were outstanding, and upon every one of these, except the Bridge Bonds, default in the payment of interest had been made. Even upon the First Mortgage Bonds about three years of overdue interest had accumulated. To this immense load of interest arrears was added a floating debt of nearly two millions of dollars, and a credit utterly ruined. To most persons it seemed hopeless to attempt saving the junior liens—the Consolidated Bonds sold at about fifteen cents upon the dollar—while the stock was regarded as entirely worthless, except, perhaps, as a means of controlling the organization.

In strong contrast to this condition of things, we now see the lowest class of bonds, created in the reorganization of 1862, established as investment securities, not inferior, in the judgment of capitalists, to the vast majority of bonds upon the market secured by first mortgages, and readily commanding in the neighborhood of par.

But the success of the stock investment is still more astonishing. Here we have the case of a road pronounced hopelessly insolvent, weighed down by series upon series of mortgage bonds, and in addition to all these by an enormous floating debt; a road upon the credit of which a barrel of oil could hardly be purchased—no one would sell the Company anything without individual and personal security, and the Chairman himself, on many occasions, was compelled to make advances to large amounts, to raise the means actually required to keep the road in operation; a Company subject to foreclosures, to judgments, executions and sales of all kinds, in which it seemed impossible that the stock interest could survive. Nevertheless it was rescued, and in 1869 the very shares issued in substitution for the foreclosed stock became equal beneficiaries in a dividend fund, sufficient to pay upon them, and upon an equally large new capital, in the meantime, put upon the road, the extraordinary rate of twelve per cent. per annum, payable quarter-yearly, free of all taxes, and this after providing for bonded interest of every kind—and in perpetuity. It will be observed, that although the present stock is a guaranteed *seven* per cent. stock, the dividend fund provided in the lease to the Pennsylvania Railroad Company was equivalent to *twelve* per cent. on the whole capital then outstanding; but the stockholders themselves, considering a seven per cent. stock more convenient and saleable, reduced this rate to seven per cent. by increasing the volume of stock upon which the fixed dividend fund of $1,380,000, annually, was distributable. By this means, each holder of the stock existing at the date of the lease became entitled to exchange every one hundred shares held by him for one hundred *and seventy-one* shares of the stock now outstanding—an arrangement which is believed to have proved very acceptable and beneficial to all.

Having indicated the results which have been achieved, a brief reference to the principal measures adopted by those in charge of the interests of the stockholders and creditors will not be inappropriate.

In the Autumn of 1860, the affairs of the Company being in the disastrous condition before referred to, some of the principal holders of the various classes of bonds, and particularly the First and Second Mortgage bondholders, held frequent conferences at the office of WINSLOW, LANIER & Co., with a view of devising some means by which the undertaking might be rescued.

One of the first steps to the union and harmonization of the conflicting interests was, to place the whole line in the hands of Hon WM. B. OGDEN, of Chicago, as Receiver; and the road was operated by him, to the satisfaction of all parties, during the foreclosure proceedings.

About the 20th of October, in that year, these gentlemen agreed upon A PLAN OF REORGANIZATION which was equitable in its provisions, and so thoroughly digested and worked up that it speedily met with general acceptance.

This Plan contemplated the appointment of five gentlemen, to purchase in, under decrees of foreclosure necessary to be obtained in the Courts of four different States, whose decisions it was indispensable to have in harmony, the whole railroad, with its equipments and appurtenances, and invested them with plenary powers in regard to the organization of a new company, to be controlled by the Mortgage Creditors; and, generally, to take such measures for the interest of the creditors as they might deem advisable. The five gentlemen appointed to the trust, and who thereupon became known as the Purchasing Committee, were JAMES F. D. LANIER, SAMUEL J. TILDEN and LEWIS H. MEYER, of New York City, SAMUEL HANNA, of Fort Wayne, Indiana, and J. EDGAR THOMSON, of Philadelphia. Of these Mr. HANNA deceased several years since, and Mr. THOMSON within the current year. The three gentlemen first named now constitute the Committee. They also constitute a majority of the Executive Committee of the Board of Directors, and have served upon that Board from the organization of the Company in 1862 to the present time.

The cardinal idea of the Plan of Reorganization was, to put the road in the hands of the mortgage creditors, and to secure a permanent control to them, but at the same time to sacrifice no *bona fide* interest of any kind. It was based upon a faith in the property and its future, and was conceived in that spirit of liberality and enlightened self-interest which perceives that in regard to all great public undertakings of this character, the harsh and literal enforcement of legal rights is, considering the complicated nature of the equities, not only unjust, but unwise. The framers of the Plan of Reorganization perceived that by a policy which sought to preserve and develop, and not to destroy, general co-operation would be secured, and existing priorities saved and even strengthened. They sought to drive no interest to the factious opposition of despair. The results of this policy we see ; and the documents set forth in this book are worthy of serious consideration by all who may be called upon to aid in resuscitating undertakings of intrinsic merit which may be suffering from the present or any future financial revulsion and depression.

The Committee desire to call attention not only to the original Plan adopted in 1860, but to the modification of that Plan adopted in 1864. The reader of the mortgage securities will observe the extraordinary care taken in them on behalf of the bond-holders. In the first place, the mortgage deeds themselves were executed by the purchasers at the foreclosure sale, and not by the new Railway Company, and that road was only conveyed to the new Company with the liens of these mortgages securely fastened upon them. But in Article Eighth of the First and Second Mortgages respectively, the precaution was taken to prohibit the issue of capital stock by the Railway Company beyond the aggregate amount of six millions and five hundred thousand dollars without the consent of the bondholders, so that the bondholders possessing the voting power could never be deprived of the absolute control. But this restriction was intended only to be precautionary, and the machinery was provided by which all classes interested might reap the benefits of the future development of business. The bondholders were permanently organized as a distinct body, capable of readjusting their relative rights and the rights of the stockholders as events might suggest. In Article Sixth, which created a sinking fund, a power of future treaty with the stockholders was reserved, and this power, taken in conjunction with Article Eighth, afforded a sufficient consideration to the two interests for any future modification of their relations.

In April, 1864, it became evident that important advantages could be secured to all classes by relaxing the restrictions which the growth of business had made no longer necessary, upon conditions looking to the permanent security of all. The bondholders were duly convened, to consider the altered circumstances. The railroad had originally been built in sections, by separate companies, and with inadequate capital. It could hardly be called a completed road, when the present Company was first organized, and although, under the judicious and economical management of the bondholding interest, its condition had been greatly improved, by the use of such earngings as were from time to time available for purposes of renewal and constructions, it was still in 1864 wholly inadequate, both in condition and equipment, to the traffic then offering and constantly increasing. It was determined, therefore, to remove the restrictions on the stock, provided the proceeds of the first new issues were expended upon the road itself and upon other conditions.

The Committee call attention to the fact that under the new programme a provision was incorporated in the First and Second Mortgages by which the Company were required to pay annually and unconditionally fixed sums sufficient in themselves to extinguish these classes of bonded indebtedness long before maturity. This circumstance, taken in connection with the fact that by the organic laws relating to the Company no new indebtedness to an amount exceeding five per cent. of the capital stock can be created without the consent of the shareholders, will indicate upon how secure a basis all interests in the undertaking now rest.

The circumstances relating to the lease of the railroad, to the Pennsylvania Railroad Company, in June, 1869, and the objects accomplished by that lease are fully treated of in the body of the book, and need not be further adverted to here.

In conclusion, the Committee congratulate the stock and bond holders upon the future now secured to this great road.

The Committee have seen it in its darkest days; they have given it their best services for years, and have every reason to be gratified with the result. The Committee take pleasure in stating that in every measure looking to the benefit of those concerned in the property they have had the faithful co-operation of their associates in the Board of Directors, and particularly the efficient and hearty co-operation of Gen'l Geo. W. Cass, who has so long and worthily filled the position of President.

Nothing remains now, but for the future managers to see that the terms of the lease are faithfully adhered to—which they no doubt will be by the lessee. The lease itself contains every provision which suggested itself to experienced counsel for the due protection and enforcement of the rights of the stockholders, and all others who have or may invest in the property.

The present compilation has been made up under the direction of the Chairman of the Committee by the law firm of Sinnott & Meyer, of New York, who had peculiar facilities for making it complete, owing to Mr. Sinnott's personal familiarity with the history of the Company, through his long continued relations to Hon. S. J. Tilden.

JAMES F. D. LANIER,
Chairman.

December 8, 1874.

STATEMENT

OF THE

PURCHASING COMMITTEE.

(*Published in October*, 1862.)

The undersigned, the Trustees and Agents who were charged with the duty of re-organizing the various interests represented by them in the Pittsburgh, Fort Wayne and Chicago Railroad, having completed the distribution of the newly-issued securities and stock, to the parties entitled to them, under the Agreement of October 20th, 1860, and having delivered the road to the newly-organized Company, submit a brief statement of their acts by which these objects have been accomplished.

The railway now known as the Pittsburgh, Fort Wayne and Chicago Railway forms a continuous line of 467 miles, from Pittsburgh to Chicago, and is owned by one Company, organized under several corporate franchises, granted by States within which portions of the work are situate. Prior to the re-organization it had become one line, owned by one Company, formed August 1, 1856, by the consolidation of three distinct lines, owned by three distinct companies, to wit: the Ohio and Pennsylvania, extending from Pittsburgh, in the State of Pennsylvania, to Crestline, in the State of Ohio, 188 miles, of which 47 are in Pennsylvania, and 141 are in Ohio; the Ohio and Indiana, extending from Crestline to Fort Wayne, in the State of Indiana, 131 miles, of which 112 are in Ohio, and 19 are in Indiana; and the Fort Wayne and Chicago, extending from Fort Wayne to Chicago, in the State of Illinois, 148 miles, of which about 126 are in Indiana, and about 22 in Illinois.

At the time of the consolidation, the Ohio and Pennsylvania, and the Ohio and Indiana portions of the road had been completed, and the Fort Wayne and Chicago portion had been commenced. But a large outlay was required to finish, equip and render adequate to the character and business of a road of the first class the portions which were in a condition for the running of trains; and the completion of the section from Fort Wayne to Chicago was essential to the general objects of the line, and to the full development of the business of all its parts. That part of the work was constructed, to a large degree, by means of temporary credits, in a period when the negotiation of bonds was less practicable than had been anticipated, and the liabilities of the old Company ultimately proved to be larger than was estimated at the time of the consolidation. It was thus, that the embarrassments of the Company arose, from the inadequacy of the capital permanently at its disposal for the undertaking, and not from any miscalculation of the utility, value or productive capacity of the work itself.

This inadequacy was attempted to be removed by the issue of $10,000,000 of Construction Bonds, secured by a general mortgage of the Consolidated Company, which bonds were intended to be substituted for the former issues, and to furnish the means to complete the work. The same object was afterwards attempted by the funding of three coupons of all the existing bonds. These expedients failed.

A litigation, growing out of differences of opinion between the parties in interest, as to the proper mode of renovating the Company, resulted in the possession by a sequestrator, appointed by one of the Courts of Pennsylvania, of that part of the road which is within that State, and the possession of all the rest of the road by a Receiver appointed by the Circuit Court of the United States for the Northern District of Ohio.

The first step towards compromise was in the union of the whole road in the possession of the Hon. William B. Ogden, of Chicago, who was agreed upon by all the parties as receiver, and appointed under an order finally entered January 17th, 1860.

The negotiations which followed, between the numerous classes of creditors, resulted in a general plan of adjustment, which the undersigned were appointed to perfect and carry into execution, with very liberal discretionary powers.

The condition of the securities was complicated and confused.

There were nine kinds of bonds secured upon the road, eight of which were upon separate portions, including the one upon the Alleghany bridge; two kinds of bonds secured upon real estate of the Company, and several kinds of bonds issued to fund the various classes of coupons. On all of these bonds, except the bridge bonds, default had been made; the arrears of interest amounted to several millions; and the principal of the earliest in the series would become due in a few years, and the others at short intervals thereafter. The floating debt amounted to nearly two millions; the road was in an extremely bad condition, and an outlay of several millions of new capital was necessary to enable it to take the position to which it was entitled by its great natural advantages.

The practical impossibility of obtaining a complete substitution of new securities for the old, and for the miscellaneous liabilities, in a mode which required the separate consent of each individual, as a necessary condition of success; an indisposition to submit to partial and unjust discriminations; defects in many of the securities themselves, which had been prepared during the infancy of the railway system in this country; the inferiority unadvoidably attaching to securities incapable of being enforced, except against separate parts of a work, the value of which so much depended upon its entireness; disabilities in respect to legal remedies, growing out of the insufficient and diverse legislation and judicial construction of four States, having jurisdiction over parts of the road, the separate sections of which had been separately mortgaged, and were usually in at least two jurisdictions— all these considerations dictated a policy at once radical and comprehensive.

That policy consisted in prosecuting to a sale the suits which had already been instituted by the bondholders, and, on the purchase of the property in behalf of the creditors, the re-organization of all the interests involved, in a manner which it was believed would be beneficial to every interest, but peculiarly so to the inferior classes of creditors and to the stockholders.

The financial theory of the new securities was, that they should be reduced to a simple and uniform classification; that the payment of the principal should be deferred to a distant period; that the interest in arrear should be funded as principal; and that somewhat more than two years of interest yet to accrue should likewise be funded, in order to enable the entire earnings, for that period, to be applied to the improvement of the road; and that, after that period, the surplus earnings, beyond paying interest on the First and Second Mortgage Bonds, which are the only absolute obligations to pay interest, should be subject to such application as might be necessary to complete the road, to put it in high condition, and to provide it with all the equipment which might be necessary.

For this purpose, it was necessary to obtain systematic and uniform legislation from the four States, enabling the new securities to be made liens upon the whole property, as an entirety,—including all future acquired property, equipment, and other personal property, the franchises of the Company, and the franchise to be a corporation, and providing a mode for rendering the latter franchise available to the purchasers at a sale under any of these liens.

All these objects are at last satisfactorily accomplished. The obtaining of the necessary legislation ; the laborious negotiations with parties so numerous, residing in this country, in England and in Germany ; the conduct of the necessary legal proceedings to consummate a sale ; the purchase of the property ; the organization of the new Company ; the reception of the old securities ; the creation of the new, and their distribution to the parties entitled to them ; the adjustment of anomalous cases, arising out of the peculiar complexity of the interests involved, have occupied a period of more than two and a half years of our industrious attention to this large and difficult trust.

The result has been, that the holders of securities of even the highest class have been more benefited than they could have been by the absolute enforcement of their legal rights, while all the inferior interests have been rescued from total extinction.

We shall not advert to the new securities further than to draw attention to some prominent and distinctive characteristics.

1. The bonds are not only secured upon the whole line of the road, but, by the force of statutes in all the States, upon future acquired property, the equipment and appurtenances—even such as are deemed in some of the States to be capable of being severed by execution from the body of the road—and upon materials, supplies and implements, in all the States except Pennsylvania.

2. They are likewise secured upon the franchises to be a corporation now possessed by the Company, which are by these statutes declared to pass to the purchasers, so as to enable them to organize anew, without the necessity of obtaining future legisla-

tion—often difficult, and sometimes practicable only through submission to unjust sacrifices.

3. They secure to their holders the right to vote in all meetings of stockholders—the First and Second Mortgage Bonds at the rate of one vote for every two hundred dollars of their par value, and the Third, at the rate of one vote for every one hundred dollars—a provision which the experience of mortgage creditors of many railroads in this country, in cases where the stock was of small or merely nominal value, has shown to be an invaluable guaranty.

4. Provision has also been made in the deeds of trust for a *quasi* organization of the bondholders to enable them to act more efficiently for the protection of their interests and their legal rights.

5. The deeds of trust, by which the bonds are secured, are made by the undersigned direct to the Trustees, John Ferguson and Samuel J. Tilden, subject to which, and in subordination to the rights created by which, the Company takes its title ; a mode of conveyancing which, on careful consideration by counsel, in view of the peculiar nature and incidents of individual and corporate ownership under the laws and constitutions of this country, is deemed to be a valuable improvement.

6. These securities have been carefully prepared (in conformity, of course, to the bondholders' agreement) to make them as effectual as possible, under the best lights of all the large experiences—practical, financial and legal—in regard to railway investments of which the last ten years have been so fruitful.

In the various legal proceedings for the sale of the road, in the Circuit Courts of the United States for the Northern District of Ohio, the Western District of Pennsylvania, the District of Indiana, and the Northern District of Illinois, the bondholders have been represented by Henry Stanberry, Esq., Messrs. Hunter & Dougherty, Messrs. Ranney, Backus & Noble, and S. J. Tilden, Esq., and the defendant by the Honorable Noah H. Swayne, now Associate Justice of the U. S. Supreme Court, and the Honorable A. G. Thurman. Mr. Tilden has also acted as the general counsel for the bondholders, and, aided by Judge Swayne, prepared the securities.

Your Committee have also had the cordial and efficient co-operation of George W. Cass, Esq., now the President of the new Company.

NEW YORK, October 1, 1862.

(Signed)   JAMES F. D. LANIER,
SAMUEL J. TILDEN,   *Trustees and*
LOUIS H. MEYER,   *Purchasing*
J. EDGAR THOMSON,   *Agents.*
SAMUEL HANNA.

ORIGINAL PLAN

OF

RE-ORGANIZATION.

(Adopted October 20, 1860.)

Whereas, the Pittsburgh, Fort Wayne and Chicago Railroad Company has become embarrassed and unable to pay the interest on the mortgages upon the respective portions of its road, which were assumed by it, and which were given by the Ohio and Pennsylvania Railroad Company, the Ohio and Indiana Railroad Company, and the Fort Wayne and Chicago Railroad Company, or upon the mortgages given by the Consolidated Company.

And whereas, bills of foreclosure, asking for the sale of the said railroad, with all its appurtenances and equipments, have been filed by certain holders of bonds of the original Companies, and the trustee of the mortgages by which such bonds are secured, in the Circuit Courts of the United States for the Northern District of Ohio, and for the Western District of Pennsylvania, and for the District of Indiana, and for the Northern District of Illinois.

And whereas, a considerable expenditure is necessary for the purpose of putting the said railroad in a condition to do business with economy and efficiency ; and arrears of interest on the said mortgages have accumulated, and the principal of some of them will fall due before the road can supply the means of paying such principal, in addition to the accruing interest.

And whereas, it is indispensable that the holders of the aforesaid bonds of each class, should unite with each other to provide the means of purchasing the said railroad at the sale which has become inevitable, as well for the purpose of protecting their own interests from sacrifice as to preserve the said railroad, with its appurtenances and equipments, as an entirety, and to secure the power, while respecting existing priorities, of saving, as far as may be, the interests of junior creditors and stockholders from absolute extinction.

And whereas, a plan of mutual adjustment has been devised, which seems likely to secure general co-operation in promoting the future business of the road, as well as in attaining the aforesaid objects; which plan is substantially as follows:

PLAN.

It is proposed that agents—five in number—shall be appointed by the bondholders subscribing an agreement for the purpose, who shall be authorized to purchase the railroad, with its appurtenances and equipments, at any sale or sales thereof; and, in case of such a purchase, that a corporation or corporations shall be organized under the laws of Pennsylvania, Ohio, Indiana and Illinois, or under the laws of some one or more of those States, and the said railroad and property vested therein ; and that stock shall be created and securities made, and the same disposed of in the manner and on the conditions hereinafter expressed.

ISSUE OF FIRST MORTGAGE BONDS.

1. Bonds of the new corporation or corporations shall be made to an aggregate not exceeding five millions and two hundred and fifty thousand dollars, which shall be secured by a first lien upon the whole line of railway from Pittsburgh to Chicago, with all its appurtenances and equipments ; shall bear interest after 1st January, 1862, at the rate of seven per cent. per annum, payable in New York, semi-annually, in six equal classes, and in such manner, that one semi-annual interest payment shall fall due on the first day of each month in every year ; and shall be redeemable on the first day of July, 1911, or, at the option of the Company, at any time or times after the first day of July, 1866.

2. These bonds shall be convertible, at the option of the holder, into bonds bearing six per cent. interest, but irredeemable except by the operation of a sinking fund herein provided, which sinking fund shall consist of one per cent. on the amount of bonds so converted, to be reserved at the same times at which interest on the said bonds shall be payable, and of all the surplus net earnings of the Company, after paying interest on its bonds.

and dividends, at the rate of six per cent., on its stock, until two and one-half millions of dollars shall have been redeemed.

3. These bonds shall entitle the holders to vote at all meetings of stockholders of the Company, at the rate of one vote for every two hundred dollars of the par amount of such bonds.

ISSUE OF SECOND MORTGAGE BONDS.

1. Bonds shall also be made, to an aggregate not exceeding five millions and one hundred thousand dollars, which shall be secured by a second lien, and be redeemable and convertible into irredeemable six per cent. bonds, in like manner as is herein provided in respect to the First Mortgage Bonds; and shall bear interest, after 1st April, 1862, at the rate of seven per cent. per annum, payable in like manner as on the First Mortgage Bonds.

2. These bonds shall entitle the holders to vote in like manner and extent as the First Mortgage Bonds.

3. Provision shall be inserted in the said mortgage, that, in case of sale by virtue thereof. a portion of the said bonds, not exceeding four hundred thousand dollars, in the aggregate, shall have priority in respect to the payment of the principal.

ISSUE OF THIRD MORTGAGE BONDS.

1. Bonds shall also be made, not exceeding two millions of dollars, in the aggregate, which shall entitle the holder, after the first day of April, 1862, to such net earnings, not exceeding seven per cent. per annum, as may have been made in each preceding year, after paying interest on prior mortgage bonds, but in priority to dividends on the stock, or any expenditure other than such as may be necessary to maintain and renew the railway, its appurtenances and equipments; the application of such earnings shall be secured by a trust deed, and the coupons shall, by reference thereto or otherwise, express the nature of the obligations.

2. These bonds shall entitle the holders to vote at all meetings of stockholders, at the rate of one vote for every one hundred dollars of the par value of such bonds.

ISSUE OF STOCK.

1. Capital stock of the new corporation or corporations may be created, to the aggregate amount of six millions and five hundred thousand dollars, but shall be limited to that amount; and the dividends thereon shall likewise be limited to the rate of six

per cent. per annum, for each six months for which such dividends may be declared.

2. Any surplus net earnings for any six months shall belong to the sinking fund, until two and one-half millions of the bonds shall have been redeemed ; and such sinking fund shall be applied to purchasing in the bonds hereinbefore mentioned, preferences in such purchases being given in the order of the priority of the liens by which the several classes of bonds may be secured, whenever and so far as such purchases can be made at or under the par value of such bonds.

FIRST MORTGAGE BONDHOLDERS.

Holders of the First Mortgage Bonds of the Ohio and Pennsylvania Railroad Company, secured on either section, holders of the First Mortgage Bonds of the Ohio and Indiana Railroad Company, and holders of 'the First Mortgage Bonds of the Fort Wayne and Chicago Railroad Company, acceding to this plan, shall be entitled to its benefits, on performing all its conditions.

1. They shall assign, as they shall be required, to such persons as may be designated for that purpose by the Purchasing Agents in the subjoined Agreement named, the bonds so held by them, and the coupons issued therewith and remaining unpaid, whether heretofore funded or not.

2. They shall thereupon become entitled to First Mortgage Bonds of the new Corporation, to be made as hereinbefore provided, equal to the par value of the bonds and coupons so assigned, except as to the fractional amounts, less than the denomination of any bonds issued, for which scrip certificates shall be given, not bearing interest until aggregated and converted into bonds.

SECOND AND THIRD MORTGAGE AND CONSTRUC-TION BONDHOLDERS.

Holders of the Second Mortgage or Income Bonds of the Ohio and Pennsylvania Railroad Company, of the Second and Third Mortgage Bonds of the Ohio and Indiana Railroad Company, and of the Construction Bonds of the Pittsburgh, Fort Wayne and Chicago Railroad Company, acceding to this plan, shall be admitted to its benefits, on performing the following conditions, viz. :

1. They shall assign, as they may be required, to such persons as may be designated by the Purchasing Agents in the subjoined

Agreement named, the bonds so held by them, and coupons issued therewith and remaining unpaid, whether funded or not.

2. They shall thereupon become entitled to Second Mortgage Bonds of the new corporation, to be made as hereinbefore provided, for principal of the bonds so assigned, and for the par amount, without interest, of such coupons of the said bonds as matured on or before the 1st of October, 1859.

3. They shall also become entitled to Third Mortgage Bonds of the new Corporation, to be made as hereinbefore provided, equal to the par amount of the coupons maturing after the 1st of October, 1859, and up to the 1st of April, 1862.

4. Holders of the Second Mortgage Bonds of the Ohio and Indiana Railroad Company shall be entitled to receive their portion of the Second Mortgage Bonds of the new Corporation, in the class in favor of which a priority of principal in the contingency mentioned is established, in consideration of the fact that the amount of the charge which they and the First Mortgage Bonds form upon the line covered by them but slightly exceeds the first lien upon the other portions of the line.

5. The foregoing provisions apply only to such bonds as were actually sold previously to the 15th day of July, 1860, and to such others as were or may be hereafter disposed of, in adjustment of the floating debt, under the authority hereinafter expressed.

REAL ESTATE BONDS.

Holders of the bonds, known as Real Estate Convertibles, issued by the Fort Wayne and Chicago Railroad Company, and payable on the first day of April, 1874, and holders of the bonds, known as Real Estate Convertibles of the Pittsburgh, Fort Wayne and Chicago Railroad Company, payable on the first day of December, 1866, may be admitted, upon assigning, to such person as may be designated for that purpose by the Purchasing Agents in the subjoined Agreement named, the bonds by them held, with the coupons issued therewith and remaining unpaid, to receive, in exchange therefor, Third Mortgage Bonds of the new Corporation.

FLOATING DEBT SECURED BY BONDS.

Construction Bonds, now outstanding as collateral, may be used in adjustment of floating debts secured by such bonds, in

the discretion of the officers of the Company, to an amount—
including such as were actually sold—not exceeding in the
aggregate 2,450 bonds. But such bonds shall not be used except
at rates which shall be first submitted to John Ferguson, and
approved by him as adequate to adjust all such cases, without
extending the aggregate herein limited.

GENERAL CREDITORS.

Holders of valid and just debts against the Pittsburgh, Fort
Wayne and Chicago Railroad Company, not included in the
preceding classes, may, upon the assignment of such debts to the
person designated for that purpose by the Purchasing Agents, be
admitted to receive an equivalent amount in the Third Mortgage
Bonds of the new Corporation ; but the nature and amount of
such debts, and, in all cases where the debts are secured by col-
laterals, the rule for the adjustment, shall be prescribed by the
Purchasing Agents, upon principles which they may deem equitable, and their decision as to the amount of the allowance, and
the terms and conditions thereof, shall be conclusive.

STOCKHOLDERS.

Holders of stock of the Pittsburgh, Fort Wayne and Chicago Railroad Company, upon the assignment of such stock
to such persons as may be designated for that purpose by the
Purchasing Agents, may be allowed to receive an equivalent
amount of stock in the new Corporation, in shares of one hundred dollars each, with scrip certificates for less amounts, not
entitling the holder to dividends.

GENERAL PROVISIONS.

1. All cases of fractional amounts shall be adjusted as hereinbefore provided with respect to the First Mortgage Bondholders.

2. The time for the performance of the conditions of this
plan shall be fixed by the Purchasing Agents of the subscribing
bondholders; or, if not so fixed by them, may be determined
by the Corporation or Corporations formed in pursuance hereof.

3. The said Purchasing Agents may act, in all cases, by a
majority of their number, and shall decide all questions which
may arise in respect to the construction or effect of any of the
provisions of this plan ; which decision, in every such case, shall

be final and conclusive. And the said Agents are hereby invested with full power and authority to execute the provisions of this plan, to supply any and every defect in any and every case which is unprovided for by its terms, and generally to do all acts and things necessary and proper, in their judgment, to carry out its objects.

4. All parties must accede to this plan within sixty days after 1st January, 1861, or they will not become entitled to its benefits without the written consent of the said Purchasing Agents.

ASSENT OF PARTIES

TO THE FOREGOING PLAN.*

---

1. AGREEMENT OF THE FIRST MORTGAGE BONDHOLDERS.

Now, therefore, each of the persons being a holder of First Mortgage Bonds, secured by first lien on the Pittsburgh, Fort Wayne and Chicago Railroad, or some part thereof, (to wit, bonds of the Ohio and Pennsylvania Railroad Company, bearing date July 1st, 1850, and secured on the section between Pittsburgh and Massilon, or bearing date January 1st, 1851, and secured on the section between Massilon and Crestline; or of the Ohio and Indiana Railroad Company, bearing date August 1st, 1852, and secured on the road between Crestline and Fort Wayne; or of the Fort Wayne and Chicago Railroad Company, bearing date July 1st, 1853, and secured on the road between Fort Wayne and Chicago), and subscribing this instrument, hereby agrees, to and with each and all the other persons, being also holders of the said bonds and subscribing this instrument, that he, the person so agreeing, hereby accepts of and consents to the terms, conditions, and provisions, applying to the aforesaid First Mortgage Bonds and the coupons belonging thereto, which are expressed in the foregoing Plan, for and in respect to the bonds and coupons held by the persons so agreeing—being the number by the said person set opposite his name, affixed hereto ; and that he will assign and deliver the bonds by him so subscribed, to such person or persons as may be designated for that purpose by the Purchasing Agents hereinafter named, or by their survivors or survivor, at such time or times, and at such place or places,

---

* The recitals are omitted, being a repetition of the Plan itself.

as may be by them or him appointed; and will accept in lieu thereof, an amount equal to the par value of the said bonds and coupons, in bonds to be made by the corporation or corporations hereafter created, and to be secured by a first lien, as in the aforesaid Plan mentioned, and with certificates for fractional amounts, as therein provided. And to that end, and for the considerations aforesaid, each of the said persons so subscribing does hereby irrevocably appoint and constitute—

J. F. D. LANIER, Esq., ⎫
L. H. MEYER, ⎬ of New York,
S. J. TILDEN, ⎭
Hon. SAM'L HANNA, of Fort Wayne,
J. EDGAR THOMSON, Esq., of Philadelphia,

to be his agents and attorneys in fact, to attend any sale or sales which may be held of the aforesaid Pittsburgh, Fort Wayne and Chicago Railroad, and of its appurtenances and equipments, or of any part or parts of the said railroad, appurtenances, and equipments, whenever and wherever the same may be offered for sale, under or by virtue of the mortgages known, respectively, as the First Mortgage of the Ohio and Pennsylvania Railroad Company, bearing date July 1st, 1850, or the First Mortgage of the said Company, bearing date January 1st, 1851, or the First Mortgage of the Ohio and Indiana Railroad Company, or the First Mortgage of the Fort Wayne and Chicago Railroad Company, or of any one or more of the said mortgages, or of any subsequent mortgage or mortgages, or at any sale or sales which shall operate to discharge in whole or in part the lien of such First Mortgages, or any of them, or any sale or sales, any proceeds of which shall be applicable upon such First Mortgages, or any of them, whether such sale or sales be pursuant to judicial decree, or to express powers; and at such sale or sales, in their or his name, or in the name of either of them, or such other name as they may deem best, to purchase the said Pittsburgh, Fort Wayne and Chicago Railroad, and its appurtenances and equipments, or any part or parts of the same, in the manner, and subject to the restrictions hereinafter expressed, to wit:

1. In case any such sale or sales shall be made separately of the portion of the said road covered by any of the said first

four mortgages, then such Agents, in bidding for and purchasing such portions, shall be deemed to act for the holders of the bonds secured by the first mortgage upon the said portion, such holders being parties hereto, and not for the holders of bonds secured on other portions of the said road ; but,

2. In case any such sale or sales shall be made jointly of portions of the said road covered by two or more of the said first mortgages, then such Agents, in bidding for, and purchasing such portions so jointly sold, shall be deemed to act for the holders of bonds secured on the portions so sold, such holders being parties hereto:

It being the intent and meaning of these presents that each subscriber hereto shall become interested in the purchase of the specific portion of the said road, upon which the bonds by him subscribed are secured, in the exact proportion which his bonds so subscribed, with the coupons belonging thereto, bear to the whole number of the bonds, with the coupons belonging thereto, which are secured thereupon ; or in the purchase of the said portion, and such other portions of the said road as may be sold jointly, in the exact proportion which his bonds so subscribed, with the coupons belonging thereto, bear to the whole number of bonds, with the coupons belonging thereto, which are secured upon the portions of the said road sold jointly, as aforesaid ; and that the said subscriber shall not be or become liable for any such bid or bids, or for any such purchase or purchases, in any manner or to any extent beyond the distributive share of the proceeds of such sale or sales to which he may be entitled, as holder of the bonds by him hereto subscribed. And for the said purposes, and in consideration as aforesaid, each person subscribing these presents hereby agrees that he will deposit with the said agents, or with John Ferguson, trustee, of the City of New York, at 35 Pine street, the bonds by him hereto subscribed, not less than thirty days previous to the earliest day which may be appointed for the sale of any portion of said road, pursuant to a judicial decree or express power ; and does hereby authorize and empower said agents to consent to any and all orders or decrees, in any court or courts, which they may deem necessary or fit, for the purpose of fixing the rule for distributing the proceeds of any sale or sales made jointly of the whole of the said railroad, or of portions thereof which are covered by more than one of the aforesaid mortgages.

And it is further agreed, that each person subscribing hereto, and holding the bonds so subscribed, may also become entitled to an interest in that portion of any purchase or purchases made in pursuance hereof, for which holders of bonds may not subscribe,—to the extent of the proportion which the bonds of such person bear to the whole number subscribed, on furnishing to the said Agents, at least thirty days before the earliest time appointed for the sale of any portion of the said road, funds to defray the cost of such interest ; *Provided*, that every such privilege shall be subject to the option of the said Agents, in their discretion, at any time or times, to allow any such non-subscribing bondholder subsequently to become a party hereto, on such terms as to them shall seem expedient ; and it is further agreed, that in case any party hereto shall decline or omit to accept the benefit of this provision, or to comply with the same, or shall fail to fulfill his obligations under this Agreement, the said Agents may hold the interest in the said purchase or purchases, to which he would otherwise become entitled, for their own account and benefit ; and in all cases every person acquiring an interest in the said purchase or purchases in place of any non-subscribing bondholders, or of any defaulting party hereto, shall be entitled to stand, as to the new securities to be issued, in the place of the party to whose position under this agreement he shall have succeeded.

And the said agents may act, in all cases, by a majority of their number, and by one or more substitutes, attorneys, or agents ; and they, or the person or persons who shall, under their authority, receive the legal title to the said road, are declared to possess, and are hereby invested with, all the legal and equitable powers, authorities, and rights of purchasers, with respect to the entire purchase or purchases which may be made in pursuance of these presents, and shall have full power and authority to convey all the estates, rights, and authority acquired by such purchaser or purchasers, to any corporation or corporations which may be formed for the purpose of holding or operating the said railroad, or any part thereof, upon such terms and conditions, and with such restrictions and agreements as to the said Agents shall seem expedient ; and generally to do all acts and things for the formation of the said corporation or corporations, and for investing them, when so formed, with the title to the railroad, and other property acquired by such purchaser or purchasers, and all acts and things for the carrying out of the objects of this agreement, provided that the aforesaid authority shall not extend to voting, in behalf of any subscriber hereto, for directors at the meeting or meetings for the organization of the said corporation or corporations.

2. AGREEMENT OF JUNIOR LIEN HOLDERS.

Now, therefore, each of the persons being a holder of bonds secured by a second or third lien on the Pittsburgh, Fort Wayne and Chicago Railroad, or some part thereof (to wit, bonds of the Ohio and Pennsylvania Railroad Company, bearing date June 30th, 1856, and secured by a second lien on the road between Pittsburgh and Crestline; bonds of the Ohio and Indiana Railroad Company, bearing date October 1st, 1853, and secured by a second lien on the road between Crestline and Fort Wayne; bonds of the Ohio and Indiana Railroad Company, bearing date September 1st, 1854, and secured by a third lien on the road between Crestline and Fort Wayne; and Construction Bonds of the Pittsburgh, Fort Wayne and Chicago Railroad Company, bearing date January 1st, 1857, and secured on the whole line between Pittsburgh and Chicago), and subscribing this instrument, hereby agrees to and with each and all other persons, being also holders of the said bonds, and subscribing this instrument, that he, the person so agreeing, hereby accepts of and consents to the terms, conditions and provisions applying to the aforesaid mortgage bonds, and the coupons belonging thereto, which are expressed in the foregoing Plan, for and in respect to the bonds and coupons held by the person so agreeing,—being the number by the said person set opposite his name, affixed hereto,—and that he will assign and deliver the bonds by him so subscribed to such person or persons as may be designated for that purpose by the Purchasing Agents hereinafter named, or by their survivors or survivor, at such time or times, and at such place or places, as may be by them or him appointed; and will accept in lieu thereof an amount equal to the par value of the said bonds and coupons, in bonds to be made by the corporation or corporations hereafter created, and to be secured by second and third liens, as in the aforesaid plan mentioned, and with certificates for fractional amounts, as therein provided; and to that end, and for the consideration aforesaid, each of the said persons so subscribing does hereby irrevocably appoint and constitute

J. F. D. Lanier, Esq.,
L. H. Meyer, } of New York,
S. J. Tilden,
Hon. Sam'l Hanna, of Fort Wayne,
J. Edgar Thomson, Esq., of Philadelphia,

to be his agents and attorneys in fact, to attend any sale or sales which may be held of the aforesaid Pittsburgh, Fort Wayne and Chicago Railroad, and of its appurtenances and equipments, or of any part or parts of the said railroad, appurtenances and equipments, whenever and wherever the same may be offered for sale, under or by virtue of the mortgages known respectively as the Mortgage of the Ohio and Pennsylvania Railroad Company, bearing date June 30th, 1856, and creating a second lien on the road between Pittsburgh and Crestline; the Second Mortgage of the Ohio and Indiana Railroad Company, bearing date October 1st, 1853; the Third Mortgage of the Ohio and Indiana Railroad Company, bearing date September 1st, 1854; and the Mortgage of the Pittsburgh, Fort Wayne and Chicago Railroad Company, bearing date January 1st, 1857; or of any one or more of the said mortgages, or at any sale or sales which shall operate to discharge, in whole or in part, the lien of such mortgages, or any of them, or any sale or sales, any proceeds of which shall be applicable upon such mortgages, or any of them, whether such sale or sales be pursuant to judicial decree or to express powers; and, in their discretion, at such sale or sales, in their or his name, or in the name of either of them, or in such other name as they may deem best, to purchase the said Pittsburgh, Fort Wayne and Chicago Railroad, and its appurtenances and equipments, or any part or parts of the same, subject to any prior liens, or discharged from any or all prior liens, as may to them seem best, but in the manner and subject to the restrictions hereinafter expressed, to wit:

1. In case any such sale or sales shall be made separately of the portion of the said road covered by any of the said mortgages, then such agents, in bidding for and purchasing such portion, shall be deemed to act for the holders of the bonds secured by

the mortgage or mortgages upon the said portion, under which such sale shall be made, and according to the respective rights of the said mortgages, such holders being parties hereto, and not for the holders of bonds secured on other portions of the road ; but,

2. In case any such sale or sales shall be made jointly of portions of the said road covered by two or more of the said mortgages, then such Agents, in bidding for and purchasing such portions so jointly sold, shall be deemed to act as aforesaid for the holders of bonds secured on the portions so sold, such holders being parties hereto :

It being the intent and meaning of these presents that each subscriber hereto shall become interested in the purchase, if made under this Agreement, of the specific portion of the said road upon which the bonds by him hereto subscribed are secured, in the exact proportion which his bonds so subscribed, with the coupons belonging thereto, bear to the whole number of the bonds, with the coupons belonging thereto, which are secured thereupon ; or in the purchase of the said portion, and such other portion of the said road as may be sold jointly, in the exact proportion which his bonds so subscribed, with the coupons belonging thereto, bear to the whole number of bonds, with the coupons belonging thereto, which are secured upon the portions of the said road sold jointly as aforesaid ; and that the said subscriber shall not be or become liable for or interested in any such bid or bids, or for or in any such purchase or purchases, in any manner or to any extent, beyond the distributive share of the proceeds of such sale or sales to which he may be entitled as holder of the bonds by him hereto subscribed.

And for the said purpose, and in consideration as aforesaid, each person subscribing these presents hereby agrees that he will deposit with the said Agents, or with John Ferguson, Trustee, of the City of New York, at 35 Pine street, the bonds by him hereto subscribed, not less than thirty days previous to the earliest day which may be appointed for the sale of any portion of said road, pursuant to a judicial decree or express power, and does hereby authorize and empower said Agents to consent to any and all orders or decrees, in any Court or Courts, which they may deem necessary or fit for the purpose of fixing the rule for distributing the proceeds of any sale or sales made jointly of the whole of the said railroad, or of portions thereof which are covered by more than one of the aforesaid mortgages.

And it is further agreed, that each person subscribing hereto, and holding the bonds so subscribed, may also become entitled to an interest in that portion of any purchase or purchases made in pursuance hereof for which holders of bonds may not subscribe, to the extent of the proportion which the bonds of such person bear to the whole number subscribed, on furnishing to the said Agents, at least thirty days before the earliest time appointed for the sale of any portion of the said road, funds to defray the cost of such interest; *Provided*, that every such privilege shall be subject to the option of the said Agents, in their discretion, at any time or times, to allow any such non-subscribing bondholder subsequently to become a party hereto, on such terms as to them shall seem expedient ; and it is further agreed, that in case any party hereto shall decline or omit to accept the benefit of this provision, or to comply with the same, or shall fail to fulfill his obligations under this Agreement, the said Agents may hold the interest in the said purchase or purchases to which he would otherwise become entitled for their own account and benefit; and in all cases, every person acquiring an interest in the said purchase or purchases in place of any non-subscribing bondholders, or of any defaulting party hereto, shall be entitled to stand, as to the new securities to be issued, in the place of the party to whose position under this Agreement he shall have succeeded.

And the said Agents may act in all cases by a majority of their number, and by one or more substitutes, attorneys, or agents ; and they, or the person or persons who shall, under their authority, receive the legal title to the said road, are declared to possess, and are hereby invested with, all the legal and equitable powers, authorities and rights of purchasers, with respect to the entire purchase or purchases which may be made in pursuance of these presents ; and shall have full power and authority to convey all the estates, rights and authority acquired by such purchaser or purchasers to any corporation or corporations which may be formed for the purpose of holding or operating the said railroad, or any part thereof, upon such terms and conditions, and with such restrictions and agreements as to the said Agents shall seem expedient ; and generally to do all acts and things for the formation of the said corporation or corporations, and for investing them, when so formed, with the title to the railroad and other property acquired by such purchaser or purchasers, and all acts and things for the carrying out of the objects of this Agreement ; provided that the aforesaid authority shall not extend to voting, in behalf of any subscriber hereto, for Directors at the meeting or meetings for the organization of the said corporation or corporations.

And it is further agreed, that in case any party hereto shall fail or omit to comply with each and every of the conditions and provisions of this Agreement, the rights of such party, by virtue hereof, may, at the option of the said Agents, be declared to be forfeited, and thereupon all interests and rights of the said party under or by virtue of this Agreement shall absolutely cease.

IN WITNESS whereof, we have hereunto set our hands, on the twentieth day of October, 1860, and have affixed opposite our names the kinds of bonds and the number of bonds by us respectively subscribed thereto.

Numbers of the Bonds.	Kind of Bonds.	Name and Residence.

[*Note.*—The foregoing plan met with general approval from the beginning, and finally with universal acquiescence.]

LAWS OF THE SEVERAL STATES

RELATING TO OR AFFECTING

THE PITTSBURGH, FORT WAYNE AND CHICAGO RAILWAY COMPANY.

I.—PENNSYLVANIA LAWS.

AN ACT to provide for the Re-Organization of the Pittsburgh, Fort Wayne and Chicago Rail Road Company.

SECTION 1. *Be it enacted by the Senate and House of Representatives of the Commonwealth of Pennsylvania, in General Assembly met, and it is hereby enacted by the authority of the same,* That in case the Rail Road of the Pittsburgh, Fort Wayne and Chicago Rail Road Company, or any part thereof, shall be sold by virtue of any mortgage or mortgages, or deed or deeds of trust, either by foreclosure or other proceedings in law or in equity, or by advertisement in pursuance of a power or authority in such mortgage or mortgages, or deed or deeds of trust, contained, the persons for or on account of whom the purchase or purchases at any such sale or sales shall have been made, or the survivors of them, shall be, and they are hereby constituted a body politic and corporate, in deed and in law; and they shall meet in the City of Pittsburgh, within thirty days after the conveyance or conveyances shall have been delivered to such purchaser, and elect Directors; notice of such meeting and election, signed by purchasers at said sale or sales, or the survivors of them, and published in two daily newspapers in said city, for five days previous to the said meeting, having been given; at which first election, all the persons for or on account of whom the said purchase or purchases shall have been made, shall be entitled to vote, in the proportion of one vote to each one hundred dollars of par value, which they may have contributed in bonds entitled to distributive shares in the proceeds of such sale or sales, or in cash, towards the said purchase or purchases; and all persons holding bonds secured by any mortgage or trust deed prior in lien to those by virtue of which such sale or sales shall have been made, shall also be entitled to one vote for every one hundred dollars of the par value of the said bonds by them respectively held; and the said election may be made by such of the aforesaid persons as may attend the same, or be represented thereat by proxy; and the said meeting may adopt such regulations and by-laws as they may deem proper for the organization of the said Corporation; and the number of Directors may then be fixed, subject to be afterwards altered at any annual meeting of the stockholders; and the Directors so elected shall continue such until the next annual meeting, the time for which may also be fixed as aforesaid; and at least three-fourths of the Directors of the said Company shall be residents of the States within which said road is located; and it shall be the duty of such Corporation, within thirty days after its organization, to make a certificate thereof, under its common seal, attested by its President and Secretary, specifying the date of such organization, and the names of its Directors, and to transmit the said certificate to the Secretary of the Commonwealth, at Harrisburgh, to be filed in his office, and a certified copy thereof shall be evidence of the existence of the said Corporation, and of its formation pursuant to this Act.

Section 2. The said Corporation, formed pursuant to this Act, shall have power to acquire, by purchase or otherwise, and to hold, use, and enjoy, the Pittsburgh, Fort Wayne and Chicago Railroad, and each and every part thereof, whether situate within or without this Commonwealth, and all equipment, machinery, tools and materials, all lands, property, franchises, rights and things connected therewith, or necessary or convenient to the use thereof, together with the tolls, income, rents, issues and profits of the same; and shall have power to maintain and operate the same as fully as might have been done by the Pittsburgh, Fort Wayne and Chicago Railroad Company, and to erect new depots, stations and other buildings, and connect the same with the said railroad, and to acquire and hold lands for the said and other needful purposes; and the said Corporation shall also possess all the faculties, powers, authorities, immunities, privileges, and franchises, at any time held by the said Pittsburgh, Fort

Wayne and Chicago Railroad Company, or by any of the corporations heretofore consolidated into the said Company, or conferred on the said Company, or the said corporations, or either of them, by any act or law of this Commonwealth, or of the States of Ohio, Indiana or Illinois; and shall have power and capacity to hold and exercise, within each and every of the said States, all the said faculties, powers, authorities, privileges and franchises, and all others which may hereafter be conferred upon it by or under any law of this Commonwealth; and to hold meetings of stockholders and Directors, and do all corporate acts, and all things, within any of the aforesaid States, as validly as it might do the same within this Commonwealth; and the said Corporation, formed pursuant to this Act, shall also have power to create and issue capital stock in shares of one hundred dollars each, and to such aggregate amount as it shall deem necessary to carry out the objects of this Act, and may establish preferences in respect to dividends, in favor of one or more classes of the said stock, in such order and manner, and to such extent, as it may deem expedient; and may confer on holders of any bonds which it may issue, or assume to pay, such rights to vote at all meetings of stockholders, not exceeding one vote for every one hundred dollars of the par amount of said bonds, as may by it be deemed advisable; which rights, when once fixed, shall attach to and pass with such bonds, under such regulations as the by-laws may prescribe, to the successive holders thereof, but shall not subject any holder to any assessment by the said Company, or to any liability for its debts, or entitle any holder to dividends; and the said Corporation may make and issue its bonds, of not less denomination than one hundred dollars each, payable at such times and places, and bearing such rates of interest, as it may deem expedient, and may hypothecate or sell such bonds, within or without this Commonwealth, at such prices as it may deem proper: *Provided, nevertheless,* that, except within six months after the organization of the said Company, *nor shall bonds be created, nor shall any debt be contracted exceeding in the aggregate at any one time five per cent., on the par amount of the capital stock of the said Company,* unless the same shall have been previously authorized by a vote of two-thirds in interest of the stock and bondholders, at a meeting duly held; and the said Corporation may secure the payment of any bonds which it may make, issue, or assume to pay, by a mortgage or mortgages, or deed or deeds

of trust, of its railroad, or of any part or parts thereof, or any
of its real estate, and may include in any such mortgage or mortgages, or deed or deeds of trust, any locomotives, cars, and other
rolling stock or equipments, and all machinery, whether then
held or thereafter to be acquired, for the constructing, operating,
repairing or replacing of the said railroad, or any part thereof,
or any of its equipments or appurtenances; all of which
property and things so included, and all fixtures or appurtenances, whether then possessed or thereafter to be acquired, shall
be subject to the lien and operation of every such mortgage or
deed of trust, and may also include all franchises held by the
said Corporation, and connected with or relating to the said railroad, and all corporate franchises of the said Company; which
said franchises are hereby declared, in case of sale by virtue of
any such mortgage or mortgages, or deed or deeds of trust, to
pass to the purchasers, so as to enable them to form a corporation in the manner herein prescribed, and to vest in such corporation all the faculties, powers, authorities, immunities, privileges and
franchises conferred by this Act; and the said Corporation may
do all things which may be necessary or convenient to carry into
full effect the powers hereby granted, and the powers hereby
conferred may be exercised by the Directors thereof.

SECTION 3. That the said Corporation, formed pursuant to this
Act, shall, within six months after its organization, likewise have
power to assume such debts, liabilities, and claims against the said
Pittsburgh, Fort Wayne and Chicago Railroad Company, and
make such settlements or adjustments with any of the stockholders, or other parties interested therein, as it may deem proper,
and for the said purposes to use such portion of the stock or bonds
hereby authorized to be created, and in such manner as it may
deem necessary.

SECTION 4. That full authority is hereby given to the corporate authorities of the several county, township, city, village, or
other municipal corporations, owning or holders of stock in the
said Company, and to all persons holding the same in any fiduciary capacity, to accept and receive, under the re-organization,
such portion of the new stock as may be apportioned to the stock
so owned and held.

JOHN M. THOMPSON,

Speaker of the House of Representatives, pro tem.

WM. M. FRANCIS,

Speaker of the Senate.

APPROVED the thirty-first day of March, Anno Domini one thousand eight hundred and sixty.

WM. F. PACKER.

OFFICE OF THE SECRETARY OF THE COMMONWEALTH, }
HARRISBURGH, April 2, 1860. {

PENNSYLVANIA, SS :

I do hereby certify that the foregoing and annexed is a full, true and correct copy of the original Act of the General Assembly, as the same remains on file in this office.

IN TESTIMONY WHEREOF, I have hereunto set my hand, [SEAL.] and caused the seal of the Secretary's Office to be affixed, on the day and year above written.

H. L. DIEFFENBACH,
Deputy Secretary of the Commonwealth.

AN ACT RELATING TO CERTAIN CORPORATIONS.

SECTION 1. *Be it enacted by the Senate and House of Representatives of the Commonwealth of Pennsylvania, in General Assembly met, and it is hereby enacted by the authority of the same,* That it shall and may be lawful for any Railroad Company created by and existing under the laws of this Commonwealth, from time to time, to purchase and hold the stock and bonds, or either, of any other Railroad Company or Companies, chartered by or of which the road or roads is or are authorized to extend into this Commonwealth ; and it shall be lawful for any Railroad Companies to enter into contracts for the use or lease of any other Railroads, upon such terms as may be agreed upon with the Company or Companies owning the same, and to run, use and operate such road or roads in accordance with such contract or lease: *Provided,* that the roads of the Companies so contracting or leasing shall be directly, or by means of intervening railroads, connected with each other.

ELISHA W. DAVIS,
Speaker of the House of Representatives.
JOHN P. PENNY,
Speaker of the Senate, pro tem.

APPROVED—the twenty-third day of April, Anno Domini one thousand eight hundred and sixty-one.

A. G. CURTIN.

A SUPPLEMENT TO AN ACT, entitled "An Act to provide for the Re-organization of the Pittsburgh, Fort Wayne and Chicago Railroad Company," approved March thirty-first, Anno Domini one thousand eight hundred and sixty, to provide for a Classification of the Board Directors.

SECTION 1. *Be it enacted by the Senate and House of Representatives of the Commonwealth of Pennsylvania, in General Assembly met, and it is hereby enacted by the authority of the same,* That it shall be lawful for the Board of Directors of the Pittsburgh, Fort Wayne and Chicago Railway Company, by lot, or otherwise, to so classify the members thereof, that one-fourth (as near as may be) shall terminate their official term as directors at the first next annual election thereafter, and one-fourth at each subsequent election; and after being thus classified, the stock and bondholders shall elect only the numbers of the Board of Directors necessary to fill the vacancies created by the expiration of the period of services fixed as aforesaid.

JOHN CLARK,
Speaker of the House of Representatives.

RUSSELL ERRETT,
Speaker of the Senate, pro tem.

APPROVED—the third day of February, Anno Domini one thousand eight hundred and sixty-nine.

JNO. W. GEARY.

AN ACT Supplementary to an Act relating to certain corporations, approved the third day of April, Anno Domini one thousand eight hundred and sixty-one.

SECTION 2. *Be it enacted by the Senate and House of Representatives of the Commonwealth of Pennsylvania, in General Assembly met, and it is hereby enacted by the authority of the same,* That it shall and may be lawful for any Railroad Company or Companies, created by or existing under the laws of this Commonwealth, from time to time, to purchase and hold the stock and bonds, or either, or to agree to purchase or guarantee the payment of the principal or interest, or either, of the bonds

of any other Railroad Company, or Companies Chartered by it
or existing under the laws of any other State.

JOHN CLARK,
Speaker of the House of Representatives.

WILLIAM WORTHINGTON,
Speaker of the Senate.

APPROVED the seventeenth day of March, Anno Domini one
thousand eight hundred and sixty-nine.

JNO. W. GEARY.

II. ILLINOIS LAWS.

AN ACT to perfect the Title of the Purchasers of the Pittsburgh, Fort Wayne and Chicago Railroad, and to enable
them to form a Corporation, and defining the Powers and
Duties of such Corporation.

SECTION 1. *Be it enacted by the People of the State of Illinois, represented in the General Assembly,* That in case the
Railroad of the Pittsburgh, Fort Wayne and Chicago Railroad
Company, or any part thereof, shall be sold by virtue of any
mortgage or mortgages, or deed or deeds of trust, either by foreclosure or other proceedings in law or equity, or in pursuance of
a power in such mortgage or mortgages or deed or deeds of trust
contained, or by the joint exercise of the said authorities, the
purchaser or purchasers of the same, or their survivors or survivor, or they, or their, or he and his associates, or their or his
assigns, may form a corporation, by filing in the office of the
Secretary of State, under their or his signature, specifying the
name of such corporation, the number of Directors, the names
of the first Directors, the period of their services, not exceeding
one year, the amount of the original capital, and the number of
shares into which such capital is to be divided ; and thereupon
the persons who shall have signed such certificate, and their successors, shall be a body politic and corporate, by the name stated
in such certificate, and a copy of such certificate, attested by the
signature of the Secretary of State, or his deputy, shall, in all

courts and places, be evidence of the due formation and existence of the said Corporation, and of the facts in the certificate stated.

Section 2. The said corporation, formed pursuant to this Act, shall have power to acquire, by purchase or otherwise, and to hold, use and enjoy, the Pittsburgh, Fort Wayne and Chicago Railroad, and each and every part thereof, whether situate within or without this State, and all equipments, machinery, tools, and materials, all lands, property, franchises, rights, and things connected therewith, or necessary or convenient to the use thereof, together with the tolls, income, rents, issues, and profits of the same, and shall have power to maintain and operate the same as fully as might have been done by the Pittsburgh, Fort Wayne and Chicago Railroad Company, and to erect new depots, stations and other buildings, and connect the same with the said Railroad, and to acquire and hold lands for the said and other needful purposes; and the said corporation shall also possess all the faculties, powers, authorities, immunities, privileges, and franchises, at any time held by the said Pittsburgh, Fort Wayne and Chicago Railroad Company, or by any of the corporations heretofore consolidated into the said Company, or conferred on the said Company, the said Corporations, or either of them, by any act or law of this State, or of either of the States of Ohio, Indiana, or Pennsylvania, and shall have power and capacity to hold and exercise within each and every of the said States, and, so far as it may deem necessary to the general objects of its business, within any other of the United States, all the said faculties, powers, authorities, privileges, and franchises, and all others which may hereafter be conferred upon it by or under any law of this State, or of any of the aforesaid States, and to hold meetings of Stockholders and Directors, and do all corporate acts, and all things, within any of the aforesaid States, as validly as it might do the same within this State, and may consolidate with any corporations of such other States authorized to hold, maintain, and operate the aforesaid railroad; and the said Corporation, formed pursuant to this act shall also have power to create and issue capital stock in shares of one hundred dollars each, and to such aggregate amount as it shall deem necessary to carry out the objects of this act; and may establish preferences in respect to dividends, in favor of one or more classes of the said stock, in such order and manner, and to such extent, and with such securities, as it may deem expedient, and may confer on holders of any bonds which it may issue or assume to pay such rights to vote at all meetings of stockholders, not exceeding one vote for every one hundred dollars of the par amount of the said bonds, as may by it be deemed advisable, which rights, when once fixed, shall attach to and pass with such bonds, under such regulations as the by-laws may prescribe, to the successive holders thereof, but shall not subject any holder to assessment by the said Company, or to any liability for its debts, or entitle any holder to dividends; and the said Corporation may make and issue its bonds, of not less denomination than one hundred dollars each, payable at such times and places, and bearing such rates of interest, as it may deem expedient, and may hypothecate or sell such bonds, within or without this State, at such prices as it may deem proper: *Provided, nevertheless,* That, except within six months after the organization of said Company, *no bond shall be created, nor shall any debt be contracted, exceeding in the aggregate, at any one time, five per cent. on the par amount of the capital stock of the said Company,* unless the same shall have been previously authorized by a vote of two-thirds in interest of the stock and bond-holders, at a meeting duly held; and the said Corporation may secure the payment of any bonds which it may make, issue, or assume to pay, by a mortgage or mortgages, or deed or deeds of trust, of its Railroad, or of any part or parts thereof, or of any other of its property, real or personal, and may include in any such mortgage or mortgages, or deed or deeds of trust, any locomotives, cars, and other rolling stock or equipments, and any machinery, tools, implements, fuel, and materials, or other real or personal estate, whether then held or thereafter to be acquired, for the constructing, operating, repairing, or replacing of the said Railroad, or any part thereof, or of any of its equipments or appurtenances; all of which property and things so included, whether then possessed or thereafter to be acquired, shall be subject to the lien and operation of every such mortgage or deed of trust, in the same manner and with the like effect as if all such property and things constituted a part of the said Railroad; and may also include all franchises held by the said Corporation and connected with or relating to the said Railroad, and all corporate franchises, which are hereby declared, in case of sale by virtue of any such mortgage or mortgages, or deed or deeds of trust, to pass to the purchaser or purchasers, so as to enable him or them to form a Corporation, in the manner herein described; and to vest in such Corporation all the faculties, powers, authorities, immunities, privileges and franchises conferred by this act. And the said Corporation may do all things which may be necessary or convenient to carry into full effect the powers hereby granted; and the powers hereby conferred may be exercised by the Directors thereof.

SECTION 3. That the said corporation, formed pursuant to this act, shall, within six months after its organization, likewise have power to assume such debts, liabilities, and claims against the said Pittsburgh, Fort Wayne and Chicago Railroad Company, and make such settlements or adjustments with any of the stockholders, or other parties interested therein, as it may deem proper; and, for the said purposes, to use such portion of the stock or bonds hereby authorized to be created, and in such manner as it may deem necessary.

SECTION 4. That in case the said Pittsburgh, Fort Wayne and Chicago Railroad, or any part thereof, shall be decreed by any Court having jurisdiction in any State, or part of a State, composing a judicial district within which a part of said Railroad is situated, to be sold by virtue of any mortgage or mortgages, or deed or deeds of trust, upon the same, containing a power of sale to the trustee or trustees, it shall be lawful for the said trustee or trustees to unite with the proper officer in making such sale, or, under the order of the Court, to make such sale at the time and place appointed by the Court, and with such notices as may be ordered by the Court; and to execute a conveyance of the said Railroad, or the part thereof which may be so sold; and such sale and conveyance shall be a valid and effectual execution of the powers of sale and conveyance contained in the said mortgage, mortages, or deed or deeds of trust, and shall operate to invest the purchasers with the title to the Railroad property and things sold as aforesaid, free and discharged from all rights and equity of redemption by the mortgagor or junior incumbrancer, or any other party whatsoever.

SECTION 5. That full authority is hereby given to the corporate authorities of the several counties, townships, cities, villages, or other municipal corporations, owning or holding stock in the said Company, and to all persons holding the same in any fiduciary capacity, to transfer, assign, or surrender the same, and to accept and receive, under the re-organization, such portion of the new stock as may be apportioned to the stock so owned and held.

SECTION 6. That this Act shall be a public Act, and shall take effect from and after its passage.

S. M. CULLOM,
Speaker of the House of Representatives.

FRANCIS A. HOFFMAN,
Speaker of the Senate.

APPROVED February eighth, eighteen hundred and sixty-one.

RICHARD YATES,
Governor.

UNITED STATES OF AMERICA, }
 State of *Illinois.* } ss :

Seal of the State of Illinois.   I, O. M. HATCH, Secretary of State of the State of Illinois, do hereby certify that the foregoing is a true copy of an Enrolled Law now on file in my office.

IN WITNESS WHEREOF, I have hereunto set my hand, and affixed the Great Seal of State, at the City of Springfield, this 14th day of February, A. D. 1861.

O. M. HATCH,
Secretary of State.

AN ACT to amend an Act entitled "An Act to provide for a general system of railroad incorporations, approved November 5, A. D. 1849.

SECTION 1. *Be it enacted by the People of the State of Illinois, represented in General Assembly,* That it shall be lawful for any railroad company, organized and doing business, or which shall hereafter be organized under any law or laws of this State, by resolution of its Board of Directors or Executive Committee, to divide its Board of Directors into three classes, numbered consecutively ; the term of office of the first class to expire on the day of the annual election of said Company then next ensuing, the second class one year thereafter, and the third class two years thereafter.

At each annual election after such classification, the stockholders of such Company shall elect, for a term of three years, a number of Directors equal to the number in the class whose term expires on the day of such election, all other vacancies to be filled in accordance with the by-laws of said Company.

SECTION 2. This act shall be a public act, and take effect and be in force from and after its passage.

Approved March 22, 1869.

III. INDIANA LAWS.

AN ACT to Legalize, Authorize, and Regulate the Sale of, and to perfect the title of purchasers of Railroads hereafter sold, or hereafter to be sold, by foreclosure, or other proceedings in law or equity, and to enable them to organize Corporations, and to exercise corporate and other powers, to provide for the payment of stock injured by such Corporations, and to provide for the payment of ticket and freight balances.

SECTION 1. *Be it enacted by the General Assembly of the State of Indiana :* That in case a majority in interest of the creditors of a Railroad Company, and a majority in interest of the Stockholders of such Company, shall agree upon a plan for the readjustment or capitalization of the debt and stock thereof, thereupon an agreement, as aforesaid, either before or after a sale of said Railroad under judicial proceedings, and a purchase at such sale by trustees, on behalf of the parties to such agreement, all the franchises and powers, including the franchises to act as a Corporation, conferred by the charter of such Railroad Company, shall pass by such sale, and vest in the said trustees, together with the Railroad and all the other property embraced in the sale ; and in case any Railroad situate wholly or partly within this State, or any part thereof situate within this State, shall, in pursuance of such agreement, be sold by virtue of any mortgage or mortgages, or deed or deeds of trust, either by foreclosure or other proceedings in law or equity, or pursuant to any power in such mortgage or mortgages or deed or deeds of trust contained, or by the joint exercise of those authorities as hereafter provided, the purchaser or purchasers of the same, or their survivor or survivors, or they, or their, or he and his associates, may form a Corporation by filing, in the office of the Secretary of State, a certificate under their or his signature, specifying the name of such Corporation, the number of directors, the names of the first directors, and the period of their service, not exceeding one year, the amount of the original capital, and the number of shares into which such capital is to be divided; and the persons signing said certificate, and their successors, shall be a body politic and corporate by the name therein specified; and a copy of such certificate, attested by the signature of the Secretary of State, or his deputy, shall, in all courts and places, be evidence of the due organization and existence of the said Corporation, and of the facts in the said certificate stated : *Provided*, that no sale under the provisions of this act shall be valid, unless notice thereof, stating time and place of sale, shall have been published in some newspaper of general circulation in the City of New York, and also by publishing said notice in at least one newspaper of general circulation, published in each county in this State through which said Railroad may run, not less than thirty nor more than sixty days, at the discretion of the Court ordering said sale, immediately preceding said sale. And all sales of Railroads, made under the order or decree of a Court of Record, are hereby legalized as fully as though the sale had been made in pursuance of this act: *Provided*, that nothing herein contained shall be construed to legalize the decree itself, or to correct any error therein, or to legalize the sale or conveyance of any real estate, by, or to any Railroad Company, or to legalize any consolidation by any Railroad Companies in this State, but only to confine the sale of the road bed, depot grounds, and such realty as is essential to the operations of the Railroad, including also the rolling stock, machinery and equipments upon the road, as embraced in the decree.

SECTION 2. Such corporation shall possess all the powers, rights, privileges, immunities, faculties, and franchises, in respect to the said Railroad, or the part thereof purchased, as aforesaid, which were possessed or enjoyed by the Corporation that owned or held the said Railroad, previous to such sale, by virtue of its charter or amendments thereto, or other laws of this State, or of any State, not inconsistent with the laws of this State, in which any part of the said Railroad is situate; and shall also have power, by agreement of the persons forming the said corporation, as aforesaid, or by a vote of a majority in interest of the stockholders, at any time within six months after the formation of the said corporation, to assume any debts or liabilities of the corporation which owned or held the said Railroad before the said sale, and in like manner, and within a like period, to make such adjustments with any stockholders of the said last mentioned corporation, as it may deem expedient, and for the said purposes to use such portions of the bonds and stock it may be authorized to create, as it may deem necessary, and in such manner as it may deem proper; and shall also have power to make and issue bonds, payable at such times and places, and bearing such rates of interest as it may deem expedient, and to sell or dispose of such bonds at such prices, and in such manner as it may deem proper; and to secure the payment of any bonds which it may make, issue, or assume to pay, by a mortgage or mortgages, or deed or deeds of trust, of its Railroad, or of any part thereof, or any other of its property, real or personal, and may include in such mortgage or mortgages, or deed or deeds of trust, any locomotives, cars, and other rolling stock and equipments, and any machinery, tools, implements, fuel, and materials, whether then held or thereafter to be acquired, for the constructing, operating, repairing, or replacing the said Railroad, or any part thereof, or any of its equipments or appurtenances, all of which property so included, whether then held or thereafter to be acquired, shall be subject to the lien and operation of such mortgage or mortgages, or deed or deeds of trust, all franchises held by the said corporation, and connected with or relating to the said Railroad, and all corporate franchises of the said Company, which said franchises, in case of sale, by virtue of any such mortgage or mortgages, or deed or deeds of trust, are hereby declared to pass to the purchasers, so as to enable them to form a corporation in the manner herein prescribed, and to vest in such corporations all the faculties, powers, authorities, immunities and franchises conferred by this act. And the said corporation shall have power to establish sinking funds for the redemption of any of its debts; and shall likewise have power to issue capital stock to such aggregate amount as it shall deem necessary, not exceeding any limitation which may be fixed by agreement with the persons forming the said Company, in the manner hereinbefore provided, and may establish preferences in respect to dividends, in favor of one or more classes of the said stock, in such order and manner, and to such extent, with securities, as it may deem expedient; and may confer on holders of any bonds which it may issue or assume to pay, such rights to vote at all meetings of stockholders, not exceeding one vote for every one hundred dollars of the par amount of the said bonds, as may by it be deemed advisable; which rights, when once fixed, shall attach to and pass with such bonds, under such regulations as the by-laws may prescribe, to the successive holders thereof, but shall not subject any holders to any assessments by the said Company, or to any liability from its debts, or entitle any holders to dividends. And the said corporation shall also have capacity to hold, and enjoy, and exercise within other States, the aforesaid faculties, powers, rights, immunities and franchises, and such others as may be conferred upon it by any law of the State, or of any other State in which any part of its Railroad may be situate, or in which it may do any part of its business, and to hold meetings of Stockholders and Directors, and do all corporate acts, and all things, without this State, as validly as it may do the same within this State.

SECTION 3. In case the part situate within this State of any Railroad, a part of which is situate in another State, shall become vested in a Corporation of such other State, and such Corporation shall also acquire a part situate in such other State, of the said Railroad, the said corporation may exercise and enjoy within this State, for the purpose of the said Railroad and its business, so far as it may be endowed by the laws of the State of its creation with capacity to do so, all the powers, rights, faculties, privileges, immunities, and franchises enumerated in section second of this act, and its mortgages or trust deeds shall operate as therein specified.

SECTION 4. Next in order of lien to the existing mortgage debt of the old road, shall stand the amounts due persons for labor performed, and wood and other such materials furnished the old Company in running the road, and damages for killing stock, and right of way; *Provided*, that all the property of said Company shall be liable for damages recovered against said Company for stock killed or injured by them or exempt from mortgages liens.

SECTION 5. So much of any Railroad as lies in this State, and is embraced in the mortgage or mortgages sought to be foreclosed, may be sold at any such sale as an entirety, and the Court making a decree or order of sale, may declare in the order where the principal office of the Railroad Company is situate within the State, and may order the sale to be made at the Court-House door of the County in which the principal office within the State is situate.

SECTION 6. In case of the sale of a Railroad, or any part thereof, as in the first section of this act mentioned, full power is hereby given to the corporate authorities of the several counties, cities, townships and other municipal corporations, holding stock in the Company, by which such Railroad was owned, and to all persons holding such stock in a fiduciary capacity, to surrender or assign such stock, and to accept and receive such new stock in any Corporation, which, after such sale, may become the owners of said Railroad, or any part thereof, as may be apportioned or given in respect to the said first-mentioned stock, under any reorganization of the ownership of the said Railroad.

SECTION 7. That no purchaser or purchasers of any Railroad shall be entitled to any rights or benefits under this act, until such purchaser or purchasers shall first assume and pay in money, or first-class or satisfactory securities, to be issued by the new corporation, formed upon the sale or transfer of any Railroad as herein provided for, as the creditor or creditors may elect, all ticket balances or back charges for freight, with interest, whether due upon account, judgment of a Court of Record, bond, note, or other instrument in writing, which the former Railroad Corporation may have owed or been in arrears for to any connecting Railroad Company operating a Railroad entirely or in part in this State.

SECTION 8. This Act may be amended or repealed at the discretion of the Legislature.

SECTION 9. It is hereby declared that an emergency exists for the immediate taking effect of this act; and, therefore, the same shall be in force from and after its passage.

JOHN R. CRAVENS,
President of the Senate.

CYRUS M. ALLEN,
Speaker of the House.

APPROVED and signed March 5th, 1861.

OLIVER P. MORTON,
Governor.

STATE OF INDIANA, *Office of the Secretary of State.* } To wit :

I hereby certify that the above is a correct and complete copy, as taken from the original enrollment, now on file in my office.

[SEAL.]   IN TESTIMONY WHEREOF, I have hereunto set my hand, and affixed the seal of the State, at the City of Indianapolis, this 6th day of March, 1861.

W. A. PEELLE,
Secretary of State.

AN ACT authorizing the classification of the board of directors of Railroad Companies. (Approved May 15, 1869.)

SECTION 1. *Be it enacted by the General Assembly of the State of Indiana,* That it shall be lawful for the board of directors of any Railway Company, whose road passes through this State into adjoining States, by lot or otherwise, to so classify the members thereof, that one-fourth (as near as may be) shall terminate their official terms as directors at the first next annual election thereafter, and one-fourth at each subsequent election ; and, after being thus classified, the stock and bondholders shall elect only the number of the Board of Directors necessary to fill the vacancies created by the expiration of the periods of services fixed as aforesaid.

SEC. 2. Whereas, an emergency exists for the immediate taking effect of this act, therefore the same shall take effect and be in force from and after its passage.

IV. OHIO LAWS.

A LAW *to Regulate the Sale of Railroads, and the Re-organization of the same.*

SECTION 1. *Be it enacted by the General Assembly of the State of Ohio,* That in case two-thirds in interest of the creditors of a Railroad Company, and two-thirds in interest of the Stockholders of such Company, shall agree, in writing, upon a plan for the readjustment or capitalization of the debt and stock thereof, then upon judicial proceedings for the sale of the road, under a mortgage or mortgages, or a deed or deeds of trust, the Court before which such proceedings may be had shall proceed to render a judgment or decree against the Company for the amount that may then be due and in arrear upon said securities, which judgment or decree shall, from its rendition, become a lien on all the property embraced in such securities, and upon all the franchises and powers of said Company, including its franchise to be and act as a corporation, conferred by the charter and the amendments to the charter of such Company ; and upon a sale had under such decree, and a purchase at such a sale by trustees on behalf of the parties to such agreement, appointed by said agreement, all the said property so bound by the judgment or decree, including all the said franchises, shall vest in said trustees ; provided that every such agreement shall provide that the unsecured debts of the Company, incurred for repairs or running expenses, shall be paid in money, or bonds of the re-organized Company as hereinafter provided, said bonds to be of the highest class issued. A copy of the terms of said agreement shall be filed in said Court before the rendition of said decree.

SEC. 2. That the said trustees shall, as soon as may be after the sale, call a meeting of the parties to the aforesaid agreement, by a notice signed by a majority of said trustees, or of their survivors, and published not less than once a week for four weeks, in a newspaper printed in the Cities of New York and Philadelphia, and in another newspaper printed in each county on the line of the said Railroad, specifying the day and place and object of such meeting ; the place of such meeting to be on the line of such railroad ; that at such meeting each of the parties to the aforesaid agreement shall be entitled to vote according to the provisions thereof, but not exceeding one vote for every fifty dollars of the par value of the debt or stock of such party, according to a list of voters, and of their respective interests, which shall be prepared by the said majority of the trustees, who are empowered to act as judges of the election; that such meeting, by a majority in interest of the persons present, in person or by proxy, shall be competent to retain or change the name of said corporation; to decide, for the time being, the amount of its capital, and the number of shares into which such capital shall be divided; to fix the number of directors and their term of office; to elect such directors, a majority of whom shall be residents of the State or States in which such Railroad is situated; and to do all things necessary or proper to reorganize said corporation. *Provided,* that any creditor shall be entitled to become a party to the agreement aforesaid, either at or at any time before the meeting in this section provided for; and that any stockholder shall be entitled to become a party to the agreement aforesaid at any time within one year after such meeting.

SEC. 3. That a certificate, under the common seal of such Corporation, specifying its name and the Railroad which it is to hold, maintain, and operate, shall be filed in the office of the Secretary of State, and that a copy of such certificate, duly authenticated by him under the great seal of this State, shall, in all courts and places, be evidence of a compliance with all the conditions and provisions of this act, and of the due re-organization, and of the existence of the said corporation.

SEC. 4. Upon such re-organization, and a conveyance by the trustees, or of such of them as shall be vested with the legal title, or their survivors, all the Railroad and other property and franchises and things purchased as aforesaid, and all the franchises, powers, faculties, privileges and immunities which were possessed or enjoyed by the original Company, or by any Company with which it had been consolidated, shall pass to and be vested in the said corporation as re-organized; and the same, and all property and things which the said re-organized corporation shall thereafter acquire, except as hereinafter provided, shall be taken, held, and disposed of for the use and benefit of the creditors and stockholders of the said corporation, who shall have be-

come such upon and after such re-organization, according to their
respective rights, but subject to the powers of the said corporation, and shall be in nowise chargeable in respect to any debt,
liability or claim of any creditor or stockholder, which subsisted
prior to the sale and re-organization herein provided for; but all
property of the original corporation not embraced in the said
sale shall, upon the re-organization, be vested in the said corporation as re-organized, in trust for all parties interested therein as
creditors, stockholders, or otherwise.

SEC. 5. The said corporation shall likewise have power,
at any time within six months after the re-organization, to
assume such debts or liabilities of the original Company, and
to make such adjustments or exchanges with any bondholders
of the original Company, and any stockholder, within one
year, as to the said corporation may seem expedient,
and may use for such purpose any bonds or stock which
it may be authorized to issue or create ; and it shall have power
to make and issue said bonds, payable at such times and places,
and bearing such rates of interest as it may deem expedient, not
exceeding seven per cent. per annum ; and to secure the payment
of any bonds which it may issue or assume to pay, by mortgages
or deeds of trust of its Railroad, or any other of its property, real
or personal, and to include therein, with its road, all its cars and
other rolling stock and equipments, and any machinery,
tools, implements, fuel, materials, and all other things, then held
or thereafter to be acquired, for the constructing, operating or
repairing said road, or for repairing or replacing any of its
equipments or appurtenances, as part and parcel of said Railroad, and as constituting with said road one property ; and to include in such mortgages or deeds of trust all franchises held by
said corporation and connected with or relating to said road.
and all other corporate franchises of said Company, all which said
franchises, including the franchise of being a corporation, in
case of sale by virtue of any such mortgage or deed of trust, or
of any judgment specified in section six, are hereby declared to
pass to the purchasers, so as to enable them to re-organize the
corporation in the manner herein provided ; and shall also have
power to issue capital stock to such aggregate amount as it may
deem proper, not exceeding any limit which may be fixed by
agreement with the trustees purchasing as aforesaid ; and may

establish preferences in respect to dividends, in favor of one or more classes of the said stock, in such order and manner as it may deem expedient, not exceeding such limit as may be fixed by agreement as aforesaid; and may, if authorized by the agreement mentioned in section one, confer on holders of any bonds which it may issue or assume to pay, such rights to vote at all meetings of stockholders, not exceeding one vote for every fifty dollars of the par amount of the said bonds, as may have been provided for in the agreement mentioned in section one, which rights, when once fixed, shall attach to and pass with such bonds, under such regulations as the by-laws may prescribe, to the successive holders thereof, but shall not subject any holder to any assessment by the said Company, or to any liability for its debts, or entitle any holder to dividends.

SEC. 6. The lien of the mortgages and deeds of trust authorized to be made by this act, shall be postponed to the lien of judgments recovered against said corporation, after its re-organization, for labor thereafter performed for it, or for materials or supplies thereafter furnished to it, or for damages for losses or injuries, thereafter suffered or sustained by the misconduct of its agents, or in any action founded on its contracts or liability as a common carrier, thereafter made or incurred.

SEC. 7. *The provisions of this act shall extend and apply also to corporations whose railroads are partly within and partly without this State; and a corporation of this State, possessing such a railroad, shall have capacity to exercise without this State all its powers, privileges, faculties and franchises; and a corporation of another State, possessing part of a railroad which is partly in such other State and partly within this State, may exercise and enjoy within this State all its powers, privileges, faculties and franchises, for the purpose of the said railroad and its business, not inconsistent with the laws of this State and the provisions of this act; and all mortgages and deeds of trust made by the said corporation upon its railroad, equipments, or other property within this State, shall operate in the same manner, and with the like effects, as is hereinbefore provided with respect to those corporations re-organized under the provisions of this Act;* PROVIDED, *that such part of the said railroad as is within this State shall be sudject to taxation, and shall be subject to all regulations of*

4

*law in the same manner as railroads of this State in like cases ;
and the corporation owning the same shall be subject to all duties
in respect thereto imposed by law, and to be sued and may sue, in
all cases, and in the same manner as a corporation of this State
might be sued or might sue.*

Sec. 8. Railroads, and other property mortgaged therewith
by said railroad companies, may, if the Court deem it expedient,
be sold without appraisement at judicial sales, under judgments
upon such mortgage; but in such case, in order to prevent sacrifices, and protect the interests of all concerned, the Court shall
fix a *minimum* sum, below which no such sale shall be made ;
and, in order to fix that amount, the Court may, if it deem it expedient to do so, refer the subject to a Master, with instructions
to take testimony and report upon the same.

Sec. 9. That in regard to bonds or stock, held by the State
of Ohio, counties, townships, cities, villages, or other municipal
corporations, or otherwise held in a fiduciary capacity, the Governor, County Commissioners, Trustees, Council, or other corporate body representing the State, municipal corporation, or persons
holding in a fiduciary capacity, as executors, administrators, guar-
dians or otherwise, shall be, and they are hereby authorized to
become parties to such agreement, and to control, exchange, or
manage said bonds or stock according to the terms of the agreement, and to take and receive new bonds or stock to be issued in
lieu of the original bonds or stock, which shall be held on the
same terms and subject to all liens which attached to said original
bonds or stock.

Sec. 10. That any association or company of persons which
has heretofore come into possession of the road and other property
of any railroad company within this State, by purchase under
judicial proceedings, and in pursuance of any agreement for the
capitalization of the debts and stock of such railroad company,
(to which a majority in interest of the creditors and of the stockholders of such railroad company were parties), and which may
now be reorganized under any law of this State, may accept of
the provisions of this act, by a vote of a majority interest of its
stockholders under such reorganization, at any meeting convened
upon notice for that purpose ; and, upon such acceptance, may re-

organize, as provided in this act; and such re-organized company shall be invested with all the powers, privileges and immunities conferred, and subject to all the restrictions imposed by this act, to the same extent as if such reorganization had been originally under this act; and a copy of the resolution of acceptance, as aforesaid, attested by the signature of the president and the seal of the corporation, shall be filed in the Office of the Secretary of State, and a copy thereof, duly authenticated by him under the great seal of this State, shall be due evidence of such acceptance; and all stockholders and creditors who have not participated in the agreement for capitalization, heretofore made, still have the right to participate in the benefits of such agreement, at or before the meeting which shall be convened upon notice for the purpose of accepting the provisions of this act, as provided in this section, in the same manner, and to the same extent, as if they had participated in and accepted of the terms of such agreement for capitalization, in the manner provided by any law under which such capitalization has heretofore taken place.

Sec. 11. This act shall take effect from and after its passage.

ED. A. PARROT,
Speaker pro tem. of the House of Representatives.

ROB'T. C. KIRK,
President of the Senate.

Passed April 11th, 1861.

SECRETARY OF STATE'S OFFICE, }
COLUMBUS, OHIO, April 16th, 1861. }

I, ADDISON P. RUSSELL, Secretary of State of the State of Ohio, do hereby certify the foregoing to be a correct copy of the original roll, now on file in this office.

{ The Great Seal }
{ of the }
{ State of Ohio. }
{ 1802. }

IN TESTIMONY WHEREOF, I do hereunto subscribe my name, and affix the Great Seal of the State of Ohio, at Columbus, on the day and year of the date hereof.

A. P. RUSSELL,
Secretary of State.

AN ACT, *Supplementary to an Act entitled "An Act to provide for the Creation and Regulation of Incorporated Companies in the State of Ohio," passed May 1st*, 1852.

SECTION 1. *Be it enacted by the General Assembly of the State of Ohio*, That the purchaser or purchasers of any railroad, situate wholly or partly within this State, which has been sold pursuant to judicial order, judgment or decree, or to such order, judgment or decree and express power, or the survivor or survivors, or the assigns of such purchaser or purchasers, may acquire the franchise to be a corporation originally vested in the company which held the said railroad prior to such sale, by grant of said company, under such terms and conditions as may be agreed upon by the directors of such company, with the consent of the stockholders owning two-thirds of the stock; which grant, being in the same form as is by law required to convey real estate, shall be effectual to pass the said franchise to the persons or company which shall have become the owners by purchase or assignment as aforesaid of such railroad: *Provided*, that no grant shall be made as aforesaid, unless provision shall be made for granting to the stockholders in the original comprny stock in the reorganized company upon equal terms with the stockholders thereof, and as shall be acceptable to the directors making such grant.

SEC. 2. This Act shall take effect from and after its passage.

JAMES R. HUBBELL,
Speaker of the House of Representatives.

P. HITCHCOCK,
President pro tem of the Senate.

APRIL 4, 1863.

Note.—Since the publication of the former collection of laws, and the execution of the lease to the Pennsylvania Railroad Company, the question has been raised in Ohio, by information in the nature of *quo warranto*, whether this Company (the Pittsburgh, Fort Wayne and Chicago Railway Company) is in fact a corporation of that State, and it has been decided that it is not, but is entitled to maintain and operate its railroad within the State by virtue of the powers and authorities granted by the other States in that behalf. The following is the Opinion of the Court:

THE STATE OF OHIO

agt.

THE PITTSBURGH, FORT WAYNE AND CHICAGO RAILWAY COMPANY, *et al.*

WELCH, *C. J.:*

This proceeding, as we understand the case, is not merely against the three defendants named upon the record, but against all the officers, stockholders, and other individuals, claiming to constitute the Pittsburgh, Fort Wayne and Chicago *Railway* Company. And we understand the information, as charging the defendants, not only with usurping and unlawfully exercising the franchise of being a corporation under and by virtue of the laws of Ohio, and as such unlawfully exercising and using the various liberties and franchises mentioned in the information, but also, with usurping the franchise of being a foreign corporation, and, as such foreign corporation, unlawfully exercising and using the same liberties and franchises, within this State.

The plea interposed stands in the names of three defendants named upon the record. In this, the persons so named say, that they are directors of the corporation, and they assert its legal existence and its full right to use the franchises in question. But they neither admit nor deny the charge that they assume to be members of the corporation, otherwise than by admitting that they assume to act as its directors. Under this state of pleading, and in the absence of evidence to the contrary, we must regard the directors as claiming to be members of the corporation, and consider their plea as a plea on behalf of all the defendants.

The claim set up by the defendants is, that they are " a corporation," created and existing under and by virtue of " the laws of the States of Ohio, Pennsylvania, Indiana, and Illinois," and, as such, authorized by said laws, to exercise and use all the said franchises and privileges. By this we do not understand, as the counsel for the State seems to do, that the defendants claim to be incorporated by the joint legislation of the States named, but that they claim to be a single organization of individuals, under the name of the Pittsburgh, Fort Wayne and Chicago *Railway* Company, to whom these States have severally granted similar corporate powers and franchises ; and they, therefore, claim to have in Ohio all the rights and powers both of a domestic and of a foreign corporation, and, as either, or both, the right to exercise and enjoy the franchises and privileges which they are charged with so usurping, namely, the franchises and privileges of owning, operating and maintaining their railroad in Ohio.

If the defendants are a corporation created by the laws of Ohio, it is admitted that they have all the rights and powers in question.

It seems, also, to be admitted in the agreed statement, and in the argument of counsel, though the contrary would appear to be asserted in the information, that the defendants are a foreign corporation, at least a corporation of the State of Pennsylvania. The questions to be decided therefore, are :

1. Is the Pittsburgh, Fort Wayne and Chicago *Railway* Company a corporation of Ohio.

2. If not such corporation, has it the right and power, as a foreign corporation, to own, operate and maintain its road in Ohio, and for that purpose to use and enjoy the privileges and franchises specified in the information.

We will consider these two questions in their order.

1. Are the defendants an Ohio corporation ?

Their claim is, that the consolidated Company, the Pittsburgh, Fort Wayne and Chicago *Railroad* Company was an Ohio corporation, and that its charter, " its franchises to be," or right of existence has passed to or become vested in the defendants, by virtue of the deed made under the Act of April 4th, 1863. Unless this act, and the deed made under it, are sufficient and effectual so to transfer or vest the charter of the consolidated Company, it is quite unnecessary to inquire whether that Company was or is a legal corporation of Ohio, and we are saved the necessity of considering the various questions, made and argued by counsel, touching the legality of the consolidation, and of the proceedings preliminary and antecedent thereto.

Assuming then, for the present—what I believe to be the fact —that the Pittsburgh, Fort Wayne and Chicago *Railroad* Company was an Ohio corporation, did its charter pass to or vest in the defendants, by virtue of the deed and act of 1863, and thus constitute the defendants, or rather thus constitute the Pittsburgh, Fort Wayne and Chicago Railway Company, an Ohio corporation ?

That a corporation can, when authorized by law so to do, transfer, sell, or convey its charter or franchise to be a corporation, and thus vest it in others, seems to be quite well settled by judicial decisions ; and we have no objections to make to this proposition of law, except it may be to the form of stating it. The real transaction in all such cases of transfer, sale or conveyance, in legal effect, is nothing more or less, and nothing other than, a surrender or abandonment of the old charter, by the corporators, and a grant *de novo* of a similar charter to the so called transferees, or purchasers ; to look upon it in any other light, and to regard the transaction as a literal transfer or sale of the charter, is to be deceived, we think, by a mere figure or form of speech. The vital part of the transaction, and that without which it would be a nullity, is the *law* under which the transfer is made. The statute authorizing the transfer, and declaring its effect, is the grant of a new charter, couched in few words, and to take effect upon condition of the surrender or abandonment of the old charter ; and the deed of transfer is to be regarded as mere evidence of the surrender or abandonment, according to our understanding of the cause cited by counsel for the defendants, in support of the doctrine of the transferability of such charter. This is the view entertained, wherever the Courts have spoken directly of the legal effect of such conveyance, and such seems to be the view taken by counsel themselves ; for they say, among other things : If the corporators ("of the old Company") saw fit, nobody would question their right to dissolve the old corporation and surrender their franchise to the State ; and no question could be made of the right of the State, by a general law, to provide for conferring it upon the purchasers of their property. And the counsel add : " *That is what in effect is done by this act* "—the act of 1863. We agree to this proposition of counsel, with a single proviso—we think with them, that " *that* is what in effect is done," provided anything is conditionally and effectually done.

In other words, the legislature of Ohio, by the act of 1863, have granted to the defendant a charter of incorporation similar to that held by the Pittsburgh, Fort Wayne and Chicago *Railroad* Company, *provided* the legislature, at the date of the act, had constitutional power to grant such a charter, *and provided,* the requirements of the act have been complied with by the parties. It matters not if we regard the charter granted as identical with the one surrendered—a *something* which really passed from the old or defunct corporation into the hands of the legislature, and thence to the new organization—there must be, at the time constitutional power in the legislature, not only to receive but also to reissue the charter. It must pass through legislative hands before it can take life in a new organization. It comes into their hands the work and offspring of the old constitution, but it goes out again, if at all, as the work and offspring of the new one, and subject to all its requirements and limitations.

By the present Constitution of Ohio, the power of the legislature to grant charters of incorporation is subjected to important limitations, which did not exist under the Constitution of 1802; one of these is, that the grant must be made by a general law; another is that the charter must be subject to alteration and revocation by the legislature; and a third is, that the grant must be made in some such form as will subject the stockholders to individual liability, to at least a certain extent, for the debts of the corporation. The claim upon the part of the State is, that the act of April 4, 1863, is in violation of these several provisions of the Constitution; or if the act will admit of a construction consistent with these provisions, then, the claim is, that the provisions and requirement of the act, taken in their proper and constitutional sense, have not been conformed to by the parties.

We have no hesitation in holding, that the act of 1863 is not liable to the objection that it is a "special act." It is a "general law," in our judgment, within the meaning of Art. 13, Sec. 2, of the Constitution. In so holding, we merely repeat, in substance, what has been heretofore decided by this Court in Cricket *v.* The State, 18 O. St. R., 9; Welker *v.* Porter, 18 O. St. R., 87.

The objection that if the defendants did thus acquire a charter, under the act of 1863, that charter would not be subject to alteration or repeal, has, in effect, been answered in what is said above. If the charter thus acquired, is to be regarded in law as identical with the charter of the re-organized Company, and not as a new charter issuing directly from the legislature, and if, in like manner, the charter of the re-organized Company is to be regarded not as a legislative grant made to it, but as a grant directly from the original Companies so consolidated, then it may be true, that the charter would be unalterable and irrevocable, and the act of 1863 be unconstitutional on that ground. But, as we have already said, such is not the law of the case, and the charter, if so vested, would remain, as other charters, granted under the present Constitution, liable to amendment and repeal by the legislature.

But the trouble in defendants' case arises when we attempt to reconcile their claim, that they are an Ohio corporation, under the act of 1863, with the third-named limitation in the Constitution —the limitation in regard to individual liability.

Under the present Constitution, the legislature are powerless to grant a charter to any such corporation, unless the grant is made in a form that will secure the individual liability of its stockholders for the debts of the corporation, at least to the amount of their stock over and above their subscription. This liability may be secured by an express provision in the act of incorporation; when it is to exceed the amount of the stock, it must be secured in that form. In the absence of any such provision in the act of incorporation, I presume this provision of the Constitution would enter into and form part of the act of incorporation, and to that extent execute itself. In either case, however, the act of incorporation—the grant of the charter— must be in some such form as will secure this liability. It must require of the *individuals*, availing themselves of its provisions, some acts *as such*, under and in pursuance of it, as will subject them individually to its provisions, or to this provision of the Constitution in regard to liability. If it fails to do this, it is simply unconstitutional and void.

The act of 1863, under which the defendants claim title, contains no improvision imposing liability upon individuals who may become stockholders under it. Whether the act, properly interpreted, does or does not require of the persons becoming incorporated under its provisions, acts or proceedings which will secure their individual liability, as stockholders, is totally immaterial to the present case; because, if it is to be interpreted as requiring such acts, namely, an *organization of individuals under the act,* such as is required by the act of April 11th, 1861, a deed to be made to and accepted by them, and a *taking* of stock by them, in the Company thus organized, then the defendants have put a wrong interpretation upon the act, and have failed to comply with its provisions. On the other hand, if they have rightly interpreted the act, then the act itself is unconstitutional and void, for the want of adequate provisions to secure the individual liability of stockholders becoming incorporated under its provisions. I presume it is not claimed, on behalf of defendants, that they have done any act by way of organization—the taking of stock, or the acceptance of the deed made under the Act of 1863—which subjects them, *as individuals*, to any liability, whatever, beyond that incurred by becoming members of the foreign company. They never organized under the Ohio act. Their organization was complete before it was passed. They took no stock under the Ohio act; their stock had already been taken under the Pennsylvania act; nor was the deed made to or accepted by them; it was made to and accepted by the corporation of which they were members. *As such corporation it had no power, by any act, whatever, to pledge the individual liability of its stockholders.* The powers of a corporation are limited to the common property and common interests of the organization. Over these, and within the scope and purpose of its organization, a *majority* of its members, acting through and by its officers and agents, can exercise dominion and control and bind its individual members. Beyond this common fund, and outside this scope, the corporation, as such, *is powerless to bind its individual members.* In some cases, it has been found very difficult to determine the exact line between what may be done by a majority of the corporators, thus acting by and through common agents, and what can only be effected by the individual consent of each and all; but no difficulty of the kind can occur in solving questions of individual liability. There, the line is strictly drawn and marked—the *contract* by which he becomes a member fixes the boundary between the interests of the stockholder and those which are embarked in the common enterprise, and thus subjected to the common control; and this contract, be it expressed or implied, must be interpreted in the light of the law as it existed at the time, and under which this organization is had. The private interests and rights of the stockholder, not by this contract, or some subsequent *individual* act of his, placed in the common fund, or subjected to the corporate control, are as completely outside the reach and power of the corporation as are the property and rights of strangers. The element of individual liability must be engrafted upon the stock by the law under which the organization is had, or the stock is taken, and by *virtue* of that organization or taking, or else by some subsequent *individual* assent of the stockholder; otherwise he stands liable for no more than the amount which, by his contract with the Company, he has agreed to contribute to the common fund.

In this view of the case, it plainly follows, that the defendants have not become members of an Ohio corporation, created under the present Constitution of the State, *for the reason that they have never subjected themselves to the individual liability which it imposes on stockholders,* and which it makes an indispensable element in the creation of all such corporations. Either the defendants have misinterpreted the Act of 1863, and wholly failed to conform to its provisions, or, if they have rightly interpreted it, as authorizing bestowment of a charter upon a foreign corporation without securing any individual liability of its stockholders, then the act itself is unconstitutional and void. In either alternative, the defendants are no legal corporation of Ohio. It is unnecessary, therefore, to inquire whether their charter as a corporation of Pennsylvania gives them authority, as such corporation, to accept an additional charter from another State, or whether, if they have such authority, it is competent for another State, not having a Constitution like ours, thus to grant them a second charter; that is, to make the grant directly to the corporation *eo nomine*, and not to the individuals composing it. If we concede both the authority to accept a second and foreign charter, and the general power of another State, in this manner, to make the grant, it is enough, for the present case, to say, that the power in question has been denied to the legislature of Ohio by her present Constitution.

II. The second general question involved, is, whether the defendants, as a foreign corporation, have the right, by the present laws of Ohio, to enjoy, exercise, and use the franchises and privileges specified in the information, other than that of being an Ohio corporation. That is to say—has the Pittsburgh, Fort Wayne and Chicago *Railway* Company, under the present laws of Ohio, accorded to it the right, to own, operate and maintain its road in and through the State, including the right to condemn and appropriate private property to its own use—the right of being a common carrier for reward, and the right to lease its road, under the act of March 19th, 1869 ? We answer this question in the affirmative, and we need perhaps add but little more. In American Bible Society *v.* Marshall, 15 O. S. R., 541, this Court held that a foreign corporation might purchase and own real estate in Ohio, when not forbidden by express legislation, or the general policy of the law. The ownership of such property implies its use by the owner, and the nature of the use is to be determined by the nature of the property itself.

There is not only no law of Ohio prohibiting the ownership and use of railroads in the State, by foreign corporations, and no public policy of the State to be contravened thereby, but there is abundant legislation directly to the contrary. The legislation brought into review by the agreed statement in this case, abundantly shows, that the policy of the State has been, and is, not only to permit, but to invite and encourage such ownership and use, and to place foreign companies, in this respect, on a perfectly equal footing with domestic companies. It would be strange were it otherwise. To invite their co-operation in works of great public concern, and then discriminate against them, in point of right to use and enjoy their property in the State, would not only be unjust to them, but unwise for the State. If any discrimination does exist, it is in regard to the power of condemning and appropriating private property to the use of the roads. In this case, we find what we construe to be an express grant of that power. The Pennsylvania act incorporating the defendants gives them power to condemn and appropriate private property. By the 7th Section of an act of April 11th, 1861, it is provided, that "a corporation of another State possessing part of a railroad which is partly in such other State, and partly within this State, may exercise and enjoy within this State all its powers, privileges, facilities and franchises, for the purpose of said railroad and its business, not inconsistent with the laws of this State and the provisions of said act." This provision clearly gives the right to condemn and appropriate private property in Ohio, to all railroad corporations of other States, which have the power of condemnation and appropriation given them in their charters of incorporation.

It follows, that a judgment of *ouster* will be entered against the defendants, as to the franchise of being a corporation of Ohio, and a judgment in their favor, as to the other franchises and privileges which they are charged with usurping.

Judgment accordingly.

Judge WEST, having been of counsel, did not sit in the case.

OPINION

OF

HON. R. P. RANNEY AND S. J. TILDEN.

ON

TAXATION OF OHIO STOCKHOLDERS.

Hon. G. W. Cass,
President P. F. W. & C. Ry. Co., New York:

Dear Sir: We have examined with much care the question submitted to us, relative to the *liability of the Stockholders* of your Company, residing in Ohio, to *taxation* upon their shares of stock, in addition to the taxes levied upon the property of the Company, situated in that State. This question is supposed to arise upon a late decision of the Supreme Court of that State, in which it was held, that the Company was not a corporation of that State, but held and operated its property *there* in virtue of its corporate organization in other States; thus, as is claimed, making such owners of stock, stockholders in a foreign corporation. If we had not been informed to the contrary, we should never have anticipated that such a claim would be advanced, nor do we think that it will be persisted in. But, be this as it may, we have no hesitation in declaring such an absurd and unjust claim totally destitute of any legal foundation in the Constitution and laws of that State. By the Constitution it is provided that, "laws shall be passed, taxing *by a uniform rule* all moneys, credits, investments in bonds, *stocks, joint stock companies*, or otherwise; and also all real and personal property, according to its true value in money;" and that "the property of corporations now existing or hereafter enacted shall forever be subject to taxation, *the same as the property of individuals.*"

These provisions require, 1st, the taxation of all property, whether belonging to individuals or corporations, at its true value in money; 2d, the same rate per cent. to be levied upon that value; and 3d, prohibit the burdening of one description of property more than another, either by valuing it at more than it is worth, taxing it twice, or levying more than the uniform rate per cent. upon it.

The tax laws of the State were generally revised in 1859. As a general proposition, the owner of any property was required to return it upon oath to the assessor of his township or wards, in order that it might be returned upon the county duplicate for taxation. This was true of investments in the stock of associations and corporations, *except* when *special provision* was made in the act for their taxation. By the 3d section (S. & C., Stat. 1440), it is provided that " no person shall be required to include " in his statement of the personal property, etc., etc.," any share " or portion of the capital stock or property of any company or " corporation, which is required to list or return its capital and " property for taxation in this State." The 16th section of the same act provided for the taxation of railroad and other companies, and requires the President, Secretary, or principal accounting officer of each of them, " whether incorporated by any " law of this State or not," to return to the respective county Auditors in the counties where their property was situated, " all " the personal property, which shall be held to include road-bed, " water and road stations, and such other realty as is necessary " to the daily running operations of the road, money and credits " of such company or corporations, within the State, at the actual " value in money." And again, in section 59, it is provided that " no person shall be required to list for taxation any certificate of " the capital stock of any company, the capital stock of which is " taxed in the name of said company."

By a supplementary act, passed May 1, 1862, the 16th section of the act of 1859, above referred to, was so far changed and repealed as to constitute the " County Auditors of the several " counties in this State, in which any railroad company now has " or hereafter may have its track and railway, or any part " thereof," * * * * " a board of appraisers and assessors " for such railroad company."

S. & S. Stat., 766, after providing for the proper organization of this board (prescribing their powers and duties in enforcing the attendance and examination of the officers of the corporation), and the apportionment of the assessment amongst the counties, towns and cities through which the road runs ; the very case of a road partly within and partly out of the State is provided for in the 8th section. It is then enacted that " When any railroad

" company has part of its road in this State and part thereof in
" any other State or States, the proper board shall take the value
" of such property, moneys and credits of such company so
" found and determined as aforesaid, and divide it in proportion
" to the length of such road in this State, bears to the whole
" length of such road, and determine the principal sum for the
" value of such road in this State accordingly, equalizing the rela-
" tive value thereof in this State, as provided in the fifth section
" of this act." If any other statutory provision was needed, to
show the status of this company and its stockholders, in respect
to taxation by the laws of Ohio, it would be found in the 7th section of the re-organization act of 1861 (S. & S. Stat., 129),
which, after conferring upon corporations of other States, possessing part of a railroad in this State, the most ample authority to
exercise all their corporate powers in the maintenance and operation of the road situated here, expressly provides, "that such part
" of the said railroad as is within this State shall be subject to
" taxation, and shall be subject to all regulation of laws in the
" *same manner* as railroads of this State in like cases."

It is thus made evident, beyond a shadow of a doubt, that
this company and its stockholders residing in Ohio, stand upon
the same ground as every other railroad company, and its stockholders, in that State; that there has not only been no attempt to
discriminate against it or them in the matter of taxation, but
that the corporation has been expressly made subject to taxation
upon all the property held by it for its stockholders, found in the
State, and the stockholders just as expressly relieved from all
obligation to return their respective interests in the stock, in the
return to be made by them. The fact that the road is owned by
a corporation of another State is not of the slightest importance.
Every Company owning the whole or any part of a road in the
State, " whether incorporated by any law of this State or not,"
" is required to list or return its capital and property for taxa-
" tion." And every owner of " any share or portion " of such
capital and property is expressly relieved from it. Any other
course would involve double taxation upon the same property,
and, we have no hesitation in saying, would be in direct conflict
with the Constitution of the State.

R. P. RANNEY.
S. J. TILDEN.

FEBRUARY 15, 1873.

THE DEEDS AND MORTGAGES.

1. TRUSTEES' AND MASTER COMMISSIONERS' DEED TO THE PURCHASING COMMITTEE.

THIS INDENTURE, made the nineteenth day of February, in the year of our Lord one thousand eight hundred and sixty-two, between JOHN FERGUSON and THOMAS E. WALKER, Trustees and Special Master Commissioners as hereinafter mentioned, of the first part, and JAMES F. D. LANIER, SAMUEL J. TILDEN, LOUIS H. MEYER, of the City and State of New York, JOHN EDGAR THOMSON, of the City of Philadelphia and the State of Pennsylvania, and SAMUEL HANNA, of the Town of Fort Wayne, and State of Indiana, of the second part.

WITNESSETH, that whereas the Pittsburgh, Fort Wayne and Chicago Railroad Company, a corporation existing under and by virtue of the laws of the States of Pennsylvania, Ohio, Indiana and Illinois, did cause to be made and delivered to the parties of the first part hereto, of the City, County and State of New York, a certain Deed, or Trust, or Mortgage, bearing date of the first day of January, A. D. 1857, whereby the said Company granted and conveyed unto the said parties of the first part thereto, their heirs and assigns, the entire railroad of the said Company from its terminus in the City of Pittsburgh to its terminus in the City of Chicago, and all the property and effects, rights and franchises in the said deed of trust or mortgage mentioned, and hereinafter particularly described, on the trusts, and for the purposes following (among others): that is to say, if the interest on the bonds therein mentioned should not be paid by the said Company when the same should fall due, and if such interest should remain in arrear for three months, then, at any time after such default, upon request of the holders of one hundred thousand dollars in amount of the said bonds, the said parties of the first part hereto should enter upon and take possession of and sell all and singular the said premises by the said deed of trust or mortgage conveyed, or transferred, or expressed, or intended so to be, first giving notice by advertisement in three principal daily newspapers printed in the Cities of New York, Albany, Boston, Philadelphia, Pittsburgh, Baltimore, Washington and Chicago, to be continued for sixty days, or such shorter time, for not less than twenty days, as the said Company might by resolution of their Board of Directors assent to, and upon such sale execute a full and legal conveyance of all and singular the premises sold.

And whereas, the said Company failed to pay the interest on the said bonds which fell due on the first days of January and July, 1859, on the first days of January and July, 1860, and on the first days of January and July, 1861, and all such interest has remained in arrear for more than three months, and is still unpaid.

And whereas, the said John Ferguson and Thomas E. Walker, grantees in trust as aforesaid, have been duly requested by holders of bonds secured by the said deed of trust or mortgage, exceeding in amount the sum of one hundred thousand dollars, to enforce and exercise the powers of entry and sale contained in and conferred upon them by the said deed of trust or mortgage.

And whereas, for the purpose of enforcing the rights of the holders of the bonds secured by the aforesaid deed of trust or mortgage, as well as of the holders of the bonds secured by the several deeds of trust or mortgages creating prior liens upon parts of the railroad and property in the said first mentioned deed of trust or mortgage described, a suit in Chancery was instituted in the Circuit Court of the United States for the Northern District of Ohio, in which Charles Moran and others were complainants, and the Pittsburgh, Fort Wayne and Chicago Railroad Company and others were defendants, and in which suit the parties hereto of the first part also became parties complainant; and auxiliary suits were also instituted between the same parties in the Circuit Courts of the United States for the Western District of Pennsylvania, for the District of Indiana, and for the Northern District of Illinois.

And whereas, at a term of the Circuit Court of the United States for the Northern District of Ohio, begun and held at the City of Cleveland, in the said district, on the 12th day of March, A. D. 1861, in and by a decree of the said Court, entered June 10th, 1861, in the aforesaid suit, it was adjudged and decreed,

5

among other things, that the said Pittsburgh, Fort Wayne and Chicago Railroad Company was in default in respect to the several installments of interest mentioned in the said decree upon the aforesaid bonds, and in respect to the several installments of interest, also mentioned in the said decree, upon bonds secured by prior liens. And it was further ordered, adjudged and decreed that the said Pittsburgh, Fort Wayne and Chicago Railroad Company should pay, on or before the second day of July then next following, the moneys so found to be due and unpaid, and that, unless the said Company should so pay the same, the property and effects, rights and franchises of the said Company in the said decree mentioned, and hereinafter particularly described, should be sold as in the said decree ordered and directed.

And whereas, decrees were also made in the said auxiliary suits then pending in the Circuit Courts of the United States for the Western District of Pennsylvania, the District of Indiana, and the Northern District of Illinois, respectively, between the same parties, in all things adopting the said decree of the Circuit Court for the Northern District of Ohio as the decree of the said several Courts respectively, in so far as respects those parts of the said railroad and property of the said Pittsburgh, Fort Wayne and Chicago Railroad Company in the said decrees respectively referred to, which are situate within the territorial jurisdiction of the said several Courts respectively.

And whereas, the said Pittsburgh, Fort Wayne and Chicago Railroad Company wholly failed to make the said payments, decreed to be made as aforesaid.

And whereas, the said Courts did further order and direct that all and singular the said property and effects, rights and franchises, should be sold by the said John Ferguson and Thomas E. Walker, the grantees in trust of the said mortgage or deed of trust, made by the said Pittsburgh, Fort Wayne and Chicago Railroad Company, bearing date on the first day of January, 1857, and in said decree mentioned, in their said capacity of Trustees, and also of Special Master Commissioners of the said Courts, and to that end did, in and by the said decrees, constitute and appoint the said John Ferguson and Thomas E. Walker Special Master Commissioners in the said suits, with full power as such, as well as in their said capacity as Trustees, to execute and carry into effect the orders of sale then made in and by the said decrees ; and further directed that such sale should be made

by public auction, at the Court-House of the said Circuit Court
of the United States for the Northern District of Ohio, in the
City of Cleveland, in the State of Ohio, to the highest bidder for
cash in hand, but not for less than the sum of five hundred thousand dollars, after advertisements of the said sale in three principal daily newspapers printed in the cities of New York, Albany, Boston, Philadelphia, Pittsburgh, Baltimore, Washington
City, and Chicago, for not less than twenty days; the said Pittsburgh, Fort Wayne and Chicago Railroad Company, by resolution of its Board of Directors, having assented to such period of
notice as found in and by the said decrees.

And whereas, in and by the said decrees, it was further
adjudged and decreed, that upon confirmation by the Court of the
said sale, so to be made as aforesaid, in pursuance thereof, the
purchaser or purchasers, upon full compliance with the condi-
itons of sale, and the orders of the Court, made or thereafter to
be made in the said cause, touching the payment of the said purchase money, should be entitled to take and hold, and the same
again to sell, or otherwise dispose of at pleasure, all of the said
property, rights, franchises and the appurtenances thereof, by the
said decree so as aforesaid ordered to be sold, by the same title by
which the same were, at any time before said sale, owned, claimed,
or held by the said Pittsburgh, Fort Wayne and Chicago Railroad Company, and by each and all of the said original railroad
companies which were consolidated into the said Pittsburgh, Fort
Wayne and Chicago Railroad Company, free and discharged
from the lien of each and all of the mortgages, in the said decree
mentioned, made by the said consolidated company, and the said
original companies, severally, and free and discharged from all
liability for any debt or debts, claim or claims, of whatsover
name or nature, in behalf of any and all persons, against said
consolidated and original companies, or any of them, and from
the claims of all persons for or on account of capital stock now
held or claimed in any of said companies; but subject, nevertheless, to the lien, if any such exists, in behalf of any vendor or
former owner thereof, upon any real estate included in said sale,
for purchase money of such real estate, not otherwise provided
to be paid, by the said decree, out of the proceeds of such sale, or
otherwise, by the former orders made in the said cause, including
the rights of way and depot grounds, and lots and premises, purchased since the pendency of the said suit by the agency of the

Receiver in the said causes, in the city and vicinity of Chicago, and subject, as to so much of the said railroad lying between the Federal street station, in Alleghany City, in the State of Pennsylvania, and the passenger depot of the Pennsylvania Railroad, in Liberty street, in Pittsburgh City, in said State, and of the other property connected therewith, as is embraced within a certain mortgage, made by the Ohio and Pennsylvania Railroad Company to Thomas Firth, of the City of Philadelphia, and Reuben Miller, Jr., of the City of Pittsburgh, bearing date May 6th, 1856, to the lien created by the said last-mentioned mortgage.

And whereas, pursuant to the said decrees, orders were duly issued by the said Courts to the said John Ferguson and Thomas E. Walker, embodying so much of the said decrees, in substance, as relates to the duties of the said Trustees and Special Master Commissioners, and directing and requiring them to carry the same into execution.

And whereas, in conformity with the said orders, the said John Ferguson and Thomas E. Walker, in their capacity as grantees in trust of the said deed of trust or mortgage made by the Pittsburgh, Fort Wayne and Chicago Railroad Company, and as Special Master Commissioners as aforesaid, caused the said property and effects, rights and franchises, hereinafter mentioned and described, to be advertised in the manner and for the period directed by the said decrees and orders of sale, respectively, to be sold as an entirety.

And whereas, the said John Ferguson and Thomas E. Walker, Trustees and Special Master Commissioners as aforesaid, and parties of the first part to these presents, in pursuance of the said deed of trust or mortgage, and of the power of sale therein contained, and of the said decrees and orders of the said Courts, did, on the 24th day of October, A. D. 1861, expose for sale at public auction, at the Court-House of the Circuit Court of the United States for the Northern District of Ohio, in the City of Cleveland, in the State of Ohio, the said railroad property and effects, rights and franchises hereinafter particularly described, at which sale the said property and effects, rights and franchises were struck off to the said parties of the second part to these presents for the sum of two millions of dollars, that being the highest and best bid therefor.

And whereas, in and by the said decree, the said Circuit

Court of the United States for the Northern District of Ohio did further order that the purchase money so bid by said purchasers be paid into the hands of William B. Ogden, in his capacity of Receiver in the said cause, to be by him held, applied, disbursed and paid over as by the said Court he should be ordered and required.

And whereas, it was made to appear to the said Court, that the said purchasers had paid to the said Receiver the said purchase money by them bid as aforesaid, in conformity with the terms of the said order.

And whereas, the said John Ferguson and Thomas E. Walker, grantees in trust and Special Master Commissioners as aforesaid, did, on the said 24th day of October, 1861, return the said order of sale, with their report of their proceedings had in pursuance thereof thereto annexed, to the said Circuit Court of the United States for the Northern District of Ohio, and have also returned the said orders of sale made in the said auxiliary causes to the said several Courts named in said orders respectively, together with their reports of their proceedings had in pursuance thereof, thereto annexed respectively.

And whereas, upon said report returned to the said Circuit Court of the United States for the Northern District of Ohio, the said Court, having inspected said report and the proceedings of said grantees in trust and Special Master Commissioners, did find, in and by a decree entered October 24th, 1861, in the said cause, that the sale and proceedings were in all things had and made according to law, and in conformity with the order and decree of the said Court in that behalf, and did approve and confirm the same ; and upon the said reports returned in the said several auxiliary causes, respectively, the said Circuit Courts for the Western District of Pennsylvania, for the District of Indiana, and for the Northern District of Illinois, did also find that the said sale and proceedings were in all things had and made according to law, and in conformity with the orders and decrees of the said several Courts, respectively, in that behalf, and did in like manner approve and confirm the same.

And whereas, it was further ordered by the said Courts, in and by the said decrees, that the said Special Master Commissioners and Trustees, in their said several capacities, should convey the said property and franchises to said purchasers upon their request, in conformity with the former orders of the said Courts in the premises.

And whereas, the said purchasers have duly requested the said Trustees and Special Master Commissioners, in their said several capacities, to convey the said property and franchises to the said purchasers, in conformity with the said orders.

Now THEREFORE, THIS INDENTURE WITNESSETH, That the said John Ferguson and Thomas E. Walker, parties of the first part to these presents, in their several capacities as Trustees and as Special Master Commissioners as aforesaid, in order to carry into effect the said sale so made as aforesaid, in pursuance of the power of sale and of the decrees and orders above mentioned, and in consider-ation of the premises, and of the sum of one dollar, to them in hand paid by the said parties of the second part hereto, the re-ceipt whereof is hereby acknowledged, have granted, bargained and sold, and by these presents do grant, bargain, sell and convey unto the said James F. D. Lanier, Samuel J. Tilden, Louis H. Meyer, John Edgar Thomson and Samuel Hanna, the parties of the second part, as joint tenants, and not as tenants in common, and to the survivors and survivor of the said parties of the second part, and to the heirs and assigns of such survivors and survivor, all and singular the railroad of the said Pittsburgh, Fort Wayne and Chicago Railroad Company, including the right of way there-for, the road-bed thereof, the superstructures of all sorts thereon, its water and other station-houses and shops, and the lands and grounds connected therewith, and all tools and implements used or provided to be used therein and in constructing and repairing cars and machinery for said road or the tracks and superstructures aforesaid, all turn-tables, all depots and buildings, and fixtures and structures of whatever name or nature, and the lands and grounds connected therewith, used or provided to be used in op-erating said road, and wherever situate, and all cars, engines and rolling stock belonging to the said Company, and all supplies of timber, lumber, iron, fuel and other things provided by said Com-pany, and by the original companies severally, which are consol-idated into the said Pittsburgh, Fort Wayne and Chicago Rail-road Company, to be used in operating said road, wherever situ-ate, by the same title by which the same were holden by said Com-pany and by the said original companies severally, together with all corporate franchises of said Company and of the said original companies severally, including the right and franchise of said several companies to be and act as a corporation. And also all and singular the property of every name and nature, and all rights, privileges and franchises conveyed by or embraced within the said mortgage or deed of trust made by the said Pittsburgh, Fort Wayne and Chicago Railroad Company to the said John Ferguson and Thomas E. Walker, which the said parties of the first part, as Trustees under and by virtue of the said deed or mortgage, might or could lawfully or rightfully sell and convey; together with all and singular the tenements, hereditaments and appurtenances thereunto belonging or in anywise appertaining, and the reversions, remainders, tolls, incomes, rents, issues and profits thereof; and also all the estate, right, title, interest, property, possession, claim and demand whatsoever, as well in law as in equity, of the said parties of the first part of, in and to the same, and any and every part thereof, with the appurtenances; subject, nevertheless, as to so much of the said railroad lying between the Federal street station in Alleghany City, in the State of Pennsylvania, and the passenger depot of the Pennsylvania Railroad, on Liberty street, in Pittsburgh City, in said State, and of the other property connected therewith as is embraced within a certain mortgage made by the Ohio and Pennsylvania Railroad Company to Thomas Firth of the City of Philadelphia, and Reuben Miller, Jr., of the City of Pittsburgh, bearing date May 6th, 1856, to the lien created by the said mortgage; and subject, also, as to the rights of way, depot grounds, lots and premises in the city and vicinity of Chicago, purchased since the pendency of the said suit, in which Charles Moran and others are complainants, and the said Pittsburgh, Fort Wayne and Chicago Railroad Company and others are defendants as aforesaid, by the agency of the Receiver in the said causes, to the lien of the vendors for the purchase money of the same; and subject, also, as to any real estate included in the aforesaid sale by the said Trustees and Master Commissioners to the lien, if any such exists, of any vendor or former owner of said real estate not otherwise provided to be paid by the decrees or orders in the said causes.

To have and to hold the same, with the appurtenances, subject, as aforesaid, unto the said parties of the second part, as joint tenants, and not as tenants in common; and to the survivors and survivor of the said parties of the second part, and to the heirs and assigns of such survivors and survivor, to the only proper use, and benefit, and behoof of the said parties of the second part, and

of the survivors and survivor of the said parties, and of the heirs
and assigns of such survivors and survivor, forever, as fully and
completely as the said parties of the first part, by virtue of the said
deed of trust or mortgage, held, and by virtue of their powers as
Trustees under the said deed, and as Special Master Commissioners,
as aforesaid, can, or could, sell and convey the said premises,
property and franchises, and not otherwise.

IN WITNESS WHEREOF, the said parties of the first part
have hereunto set their hands and seals the day and
year first above written.

JOHN FERGUSON. [L. s.]

Trustee and Special Master Commissioner.

THOMAS E. WALKER. [L.' s.]

Trustee and Special Master Commissioner.

Signed, sealed and delivered }
in the presence of—

JOHN RANKIN, Jr., and
JAMES FARRELLY.

STATE OF NEW YORK, }
City and County of New York. } ss:

Be it remembered, that on this 18th day of February, in the
year one thousand eight hundred and sixty-two, before me, John
A. Weeks, a Commissioner, resident in the City of New York,
duly commissioned and qualified by the executive authority, and
under the laws of the State of Pennsylvania, to take the acknowledgment of deeds, &c., to be used or recorded therein, personally
appeared the above named John Ferguson and Thomas E. Walker,
severally personally known to me, and in due form of law severally acknowledged the above indenture to be their act and deed
respectively, and severally desired that the same might be recorded as such.

IN TESTIMONY WEHREOF, I have hereunto set my hand and
affixed my official seal, the day and year aforesaid.

[SEAL.]   JOHN A WEEKS,
*Commissioner of Pennsylvania for
State of New York.*

STATE OF NEW YORK, ⎱ ss:
City and County of New York. ⎰

Be it remembered, that on the 18th day of February, in the year one thousand eight hundred and sixty-two, before me, the un dersigned, John A. Weeks, a Commissioner, resident in the City of New York, duly commissicned and qualified by the executive authority, and under the laws of the State of Ohio, to take the acknowledgment of deeds, &c., to be used or recorded therein, personally appeared John Ferguson and Thomas E. Walker, above named, and severally acknowledged the signing and seal ing of the foregoing conveyance to be their voluntary act and deed.

> IN WITNESS WHEREOF, I have hereunto set my hand and affixed my official seal, the day and year aforesaid.

[SEAL.]   JOHN A. WEEKS,
Commissioner of Ohio for State of New York.

STATE OF NEW YORK, ⎱ ss:
City and County of New York. ⎰

Be it remembered, that on the 18th day of February, in the year one thousand eight hundred and sixty-two, before me, the undersigned, John A. Weeks, a Commissioner, resident in the City of New York, duly commissioned and qualified by the execu. tive authority, and under the laws of the State of Indiana, to take the acknowledgment of deeds, &c., to be used or recorded there- in, personally appeared John Ferguson and Thomas E. Walker, the grantors in the foregoing deed, and severally acknowledged the execution of the same.

> IN WITNESS WHEREOF, I have hereunto set my hand and affixed my official seal the day and year aforesaid.

[SEAL.]   JOHN A. WEEKS,
Commissioner of Indiana for State of New York.

STATE OF NEW YORK, ⎱ ss:
City and County of New York. ⎰

Be it remembered, that on this 18th day of February, in the year one thousand eight hundred and sixty-two, before me, the subscriber, George H. Forster, a Notary Public for the State of New York, residing in the City of New York, duly commissioned and sworn, personally appeared, in the said City of New York, John Ferguson and Thomas E. Walker, whose signatures appear to the foregoing deed, and who are severally personally known to me to be the real and same persons described in and who executed the foregoing indenture, and whose names are subscribed to said deed as having executed the same, and severally acknowledged the same to be their free act and deed, and that they executed the same.

IN WITNESS, I have hereunto set my hand and affixed my official seal the day and year aforesaid.

[SEAL.]
                    GEO. HENRY FORSTER,
                                Notary Public.

STATE OF NEW YORK, } ss:
City and County of New York.

I, HENRY W. GENET, Clerk of the City and County of New York, and also Clerk of the Supreme Court for the said city and county, the same being a Court of Record, do hereby certify, that Geo. Henry Forster, whose name is subscribed to the certificate of the proof or acknowledgment of the annexed instrument, and thereon written, was, at the time of taking such proof or acknowledgement, a Notary Public in and for the State of New York, dwelling in the said city, commissioned and sworn, and duly authorized to take the same; and further, that I am well acquainted with the handwriting of such Notary Public, and verily believe that the signature to the said certificate of proof or acknowledgement is genuine.

I further certify that said instrument is executed and acknowledged according to the law of the State of New York.

IN TESTIMONY WHEREOF, I have hereunto set my hand and affixed the seal of the said Court and County, the 24th day of February, 1862.

[SEAL.]
                             H. W. GENET,
                                   Clerk.

2.—DEED OF FORMER RAILROAD COMPANY, PURSUANT TO ORDER OF COURT.

THIS INDENTURE, made this twenty-fifth day of February, in the year of our Lord one thousand eight hundred and sixty-two, by and between the PITTSBURGH, FORT WAYNE AND CHICAGO RAILROAD COMPANY, of the first part, and JAMES F. D. LANIER, SAMUEL J. TILDEN, LOUIS H. MEYER, of the City and State of New York, JOHN EDGAR THOMSON, of the City of Philadelphia and State of Pennsylvania, and SAMUEL HANNA, of the town of Fort Wayne and State of Indiana, of the second part.

WITNESSETH, whereas, in a cause in Chancery, depending in the Circuit Court of the United States for the Northern District of Ohio, wherein Charles Moran and others are complainants, and The Ohio and the Pennsylvania Railroad Company and others are defendants, the said Court ordered and decreed that the Special Master, Commissioners and Trustees named in said order and decree should sell, in the manner therein provided, the following described property and effects, to wit: The railroad of the Pittsburgh, Fort Wayne and Chicago Railroad Company, including the right of way therefor, the road thereof, the superstructure of all sorts thereon, its water and other station-houses and shops, and the lands and grounds connected therewith, and all tools and implements used or provided to be used therein, and in constructing and repairing cars and machinery for such road, or the track and superstructures aforesaid; all turn-tables, all depots and buildings, and fixtures, and structures, of whatever name or nature, and the lands and grounds connected therewith, used or provided to be used in operating said road and belonging thereto, and wherever situate, and all cars, engines and rolling stock belonging to said Company, and all supplies of timber, lumber, iron, fuel and every other thing provided by said Company, or by the several original companies which were consolidated into said Pittsburgh, Fort Wayne and Chicago Railroad Company, to be used in operating said road, wherever situate, by the same title by which the same are holden by said Company, or by said original companies severally, together with all corporate franchises of said Company, and of the said original companies severally, including the right and franchise of said several companies, to be an act, as a corporation, to be sold as an entirety; and

Whereas, The Circuit Court of the United States for the Western District of Pennsylvania, the Circuit Court of the United States for the District of Indiana, and the Circuit Court of the United States for the Northern District of Illinois, in causes pending in said Courts respectively, which were and are auxiliary to said cause first above mentioned, adopted said order and decree above-mentioned, as to so much of said premises as are situate in said districts respectively, and did order the same to be sold by the same persons, and at the same time and place, and in the same manner, as was provided by said decree first above mentioned, in respect of the whole of said premises ; and

Whereas, The said decrees, and each thereof, provided further, that upon full payment of the purchase money, and compliance with the terms of their purchase by said purchasers, to be found by the further order of the Court to be made in the premises, the said Special Masters Commissioners and Trustees, in their several capacities, or the survivor of them, shall by deed, to be executed in such form as may be proper to convey real estate in the said several States of Pennsylvania, Ohio, Indiana and Illinois, convey the premises to be sold to the purchasers, their heirs, successors or assigns, and thereupon, also, the said Pittsburgh, Fort Wayne and Chicago Railroad Company shall, by like deed, convey and confirm said premises to said purchasers, their heirs, successors or assigns ; and

Whereas, All said premises above described, under and pursuant to said decrees, and in conformity thereto, were on the 24th day of October, A. D. 1861, sold by said Trustees and Special Master Commissioners to the said James F. D. Lanier, Samuel J. Tilden, Louis H. Meyer, John Edgar Thomson and Samuel Hanna ; and

Whereas, it has been found by the further order of said Circuit Court of the United States for the Northern District of Ohio, made in said cause first above mentioned, that said purchasers have made full payment of the purchase money of said premises, and in all things complied with terms of the said purchase :

Now, in obedience to the requirement of said orders and decrees, and in consideration of the sum of one dollar paid by the parties of the second part to the said party of the first part, receipt whereof is hereby acknowledged, the said Pittsburgh, Fort Wayne and Chicago Railroad Company, party of the first part, as aforesaid, hath granted, bargained and sold, and doth hereby grant, bargain and sell, convey and confirm unto the said James F. D. Lanier, Samuel J. Tilden, Louis H. Meyer, John Edgar Thomson and Samuel Hanna, and to their heirs and assigns, all and singular the premises, property and effects aforesaid of every kind and description, and of every name and nature whatsoever, so sold as aforesaid to the said parties, of the second part, as aforesaid, to have and to hold the same to the only and proper use, benefit and behoof of them, the said parties of the second part, and to the survivor and survivors of them, as joint tenants, and not as tenants in common, to their heirs and assigns of such survivor for ever.

In witness whereof, the President of the said Pittsburgh, Fort Wayne and Chicago Railroad Company, party of the first part as aforesaid, pursuant to an order of said Company to that purpose duly made and recorded on the 24th day of October, in the year of our Lord one thousand eight hundred and sixty-one, hath hereunto subscribed his name officially, and caused the corporate seal of the said Company to be hereunto affixed the day and year first above written.

G. W. CASS,

{ Seal Railroad Company. } *President Pittsburgh, Fort Wayne and Chicago Railroad Co.*

W. H. Barnes,
Secretary.

Signed, sealed, acknowledged and } delivered in presence of

J. P. Henderson and
Th. D. Messler.

3.—THE FIRST MORTGAGE.

This Indenture, made this first day of March, in the year of our Lord one thousand eight hundred and sixty-two, between James F. D. Lanier, Samuel J. Tilden and Louis H. Meyer, of the City and State of New York, J. Edgar Thomson, of the City of Philadelphia and State of Pennsylvania, and Samuel Hanna, of the town of Fort Wayne and State of Indiana, of the first part and John Ferguson and Samuel J. Tilden, of the City and State of New York, of the second part, witnesseth:

Whereas, the Pittsburgh, Fort Wayne and Chicago Railway Company is vested with franchises to be a corporation, granted to the said Company by the States of Pennsylvania, Illinois and Indiana, respectively, and has become duly organized as a corporation, in conformity to the provisions of the said grants, with capacity, in its corporate character, to take, hold and exercise other franchises, and, particularly, with capacity to acquire, hold, maintain and operate the continuous railway extending from Pittsburgh, in the State of Pennsylvania, to Chicago, in the State of Illinois, commonly known as the Pittsburgh, Fort Wayne and Chicago Railroad, together with its equipments and appurtenances;

And whereas, the said Company has agreed with the parties of the first part to buy the aforesaid railway, and, in evidence of a portion of the consideration for the same, has made and delivered to the parties of the first part its bonds, amounting in the aggregate to the sum of five millions and two hundred and fifty thousand dollars, all of which bonds bear date on the first day of March, in the year one thousand eight hundred and sixty-two, and are payable at the office or agency of the said Company, in the City of New York, upon the first day of July, in the year one thousand nine hundred and twelve, and are redeemable, at the option of the Company, at any time after the first day of July, one thousand eight hundred and sixty-seven, on any day on which a half-yearly instalment of interest shall fall due, and are convertible, at the option of the holders thereof, upon any such day, into bonds, to be issued and secured in the same manner as the said bonds, but bearing interest at the rate of six per cent., and irredeemable, except by a sinking fund of one per cent., per annum, on the whole amount of the said six per cent. bonds which shall have been issued, to be reserved and applied in the manner hereinafter specified; all of which bonds originally issued, as aforesaid, bear interest at the rate of seven per centum, per annum, payable semi-annually at the office or agency of the said Company, in the City of New York; of which bonds six hundred and fifty, numbered, consecutively, from 1 to 650, inclusively, are each of one thousand dollars, and four hundred and fifty, numbered, consecutively, from 651 to 1,100, inclusively, are each for five hundred dollars; and the interest on all of which said eleven hundred bonds is payable on the first days of January and July, in each year; six hundred and fifty, numbered, consecutively, from 1,101 to 1,750, inclusively, are each for one thousand dollars, and four hundred and fifty, numbered, consecutively, from 1,751 to 2,200, inclusively, are each for five hundred dollars ; and the interest on which said eleven hundred bonds is payable on the first days of February and August, in each year ; six hundred and fifty, numbered, consecutively, from 2,201 to 2,850, inclusively, are each for one thousand dollars, and four hundred and fifty, numbered, consecutively, from 2,851 to 3,300, inclusively, are each for five hundred dollars, and the interest on all of which said eleven hundred bonds is payable on the first days of March and September, in each year; six hundred and fifty, numbered, consecutively, from 3,301 to 3,950, inclusively, are each for one thousand dollars, and four hundred and fifty, numbered, consecutively, from 3,951 to 4,400, inclusively, are each for five hundred dollars, and the interest on all of which said eleven hundred bonds is payable on the first days of April and October, in each year; six hundred and fifty, numbered, consecutively, from 4,401 to 5,050, inclusively, are each for one thousand dollars, and four hundred and fifty, numbered, consecutively, from 5,051 to 5,500, inclusively, are each for five hundred dollars, and the interest on all of which said eleven hundred bonds is payable on the first days of May and November, in each year ; six hundred and fifty, numbered, consecutively, from 5,501 to 6,150, inclusively, are each for one thousand dollars, and four hundred and fifty, numbered, consecutively, from 6,151 to 6,600, inclusively, are each for five hundred dollars : and the interest on all of which said eleven hundred bonds is payable on the first days of June and December, in each year :—all of which six thousand and six hundred bonds are in the form following

No.   UNITED STATES OF AMERICA. $

STATES OF PENNSYLVANIA, OHIO, INDIANA AND ILLINOIS.

PITTSBURGH, FORT WAYNE AND CHICAGO RAILWAY COMPANY.

FIRST MORTGAGE BOND.

Know all men by these presents, that the Pittsburgh, Fort Wayne and Chicago Railway Company are indebted to John Ferguson and Samuel J. Tilden, of the City of New York, or bearer, in the sum of   dollars, lawful money of the United States of America, which the said Company promises to pay to the said John Ferguson and Samuel J. Tilden, or to the bearer hereof, on the first day of July, in the year one thousand nine hundred and twelve, at the office or agency of the said Company, in the City of New York, with the interest thereon, at the rate of seven per centum per annum, payable semi-annually, at the said office or agency in the City of New York, on the first days of   and   in each year, on the presentation and surrender of the annexed coupons, as they severally become due; and in case of the non-payment of any half-yearly instalment of interest, which shall have become payable, and shall have been demanded; if such default shall continue for three months after the maturity of the said instalment, the principal of this bond shall become due in the manner and with the effect provided in the deed of trust hereinafter mentioned. But it is hereby provided, that, at any time after the first day of July, one thousand eight hundred and sixty-seven, the said Company, on any day on which a half-yearly instalment of interest shall fall due, may, at their option, redeem at par the principal of this bond. And it is further agreed, that this bond is convertible, at the option of the holder, upon any day upon which any such instalment of interest shall become payable, into a bond, to be issued and secured in the same manner as this bond, but bearing interest at the rate of six per cent. and irredeemable except by a sinking fund of one per cent., per annum, on the whole amount of the said six per cent. bonds which shall have been issued, to be reserved and applied in the manner specified in the said deed of trust.

"This bond is one of a series of eleven hundred bonds: six hundred and fifty for one thousand dollars each, and four hundred and fifty for five hundred dollars each ; which, with five other series of the like number and denominations, form an issue of sixty-six hundred bonds, numbered from one to sixty-six hundred, inclusively, and amounting in the aggregate to five millions and two hundred and fifty thousand dollars, all bearing date on the first of March, 1862, and all of like tenor, except that the coupons of the six several series mature on the first days of the six successive months of each half year ; and the payment of all of which bonds is secured by a deed of trust dated the first day of March, 1862, duly executed and delivered by James F. D. Lanier, Samuel J. Tilden, Louis H. Meyer, J. Edgar Thomson, and Samuel Hanna, to John Ferguson and Samuel J.

Tilden, trustees, and conveying the Pittsburgh, Fort Wayne and Chicago Railway, and the equipments, appurtenances and things therein described.

" And this bond is also entitled to the benefits of the sinking fund by the said deed of trust provided.

" The person appearing on the voting Bond Register of the said Company as the holder of this bond, at the time of any meeting of the stockholders of the said Company, will be entitled to one vote at such meeting for every two hundred dollars of the par amount hereof. The right to vote upon this bond is transferable on the written order of the person last registered as its holder, or on the production of the bond by the holder.

" It is agreed between the said Company and the holder of this bond, that no recourse shall be had for its payment to the individual liability of any stockholder of the said Company; and that in case of any default in the payment hereof, the said Company hereby waives the benefit of any extension, stay, or appraisement laws now existing, or that may hereafter exist.

This bond shall pass by delivery, or by transfer on the books of the Company in the City of New York. After a registration of ownership, certified hereon by the transfer agent of the Company, no transfer, except on the books of the Company, shall be valid, unless the last transfer be to bearer, which shall restore transferability by delivery. But this bond shall continue subject to successive registrations and transfers to bearer as aforesaid, at the option of each holder.

This bond shall not become obligatory until it shall have been authenticated by a certificate, endorsed hereon, and duly signed by the Trustees.

> " In witness whereof, the said Company have caused their corporate seal to be hereto affixed, and the same to be attested by the signatures of their President and Secretary, and have also caused the coupons hereto annexed to be signed by their Secretary, on this first day of March, in the year one thousand eight hundred and sixty-two.
>
> " *President.*
>
> " *Secretary.*"

And whereas, provision is herein made for the issue of bonds, bearing interest at the rate of six per centum per annum, but irredeemable, except by a special sinking fund herein established, into which the bonds hereinbefore described may be converted, at the option of the respective holders thereof, in the manner and on the conditions herein prescribed.

And whereas, the intention of these presents is, and is hereby declared to be, that all of the said bonds, whether of the original seven per cent. issue, or of the subsequent six per cent. issue, for the purpose of conversion as aforesaid, shall be equally secured by these presents, in proportion to the amount of the principal thereof unpaid, with the interest on the said principal accrued and unpaid, without discrimination or preference with reference to the times of the actual issue of the said bonds, or of the maturing of the principal thereof, or the maturing of any interest which shall have accrued thereon :

Now this Indenture witnesseth, that the parties of the first part, in consideration of the premises, and of one dollar to them in hand paid, the receipt whereof is hereby acknowledged, and in order to secure the payment of the principal and interest of the bonds aforesaid, issued or to be issued, as herein recited and provided, and every part of the said principal and interest, as the same shall become payable, according to the tenor of the said bonds, and of the coupons thereto annexed, have granted, bargained and sold, and do by these presents grant, bargain, sell, convey and transfer, unto the parties of the second part, all the right, title and interest of them, the parties of the first part, and of any or either of the said parties, acquired by virtue of a deed bearing date the eighteenth day of February, 1862, and made to the parties hereto of the first part by John Ferguson and Thomas E. Walker, both of the City and State of New York, trustees and Special Master Commissioners, in pursuance of decrees of the Circuit Courts of the United States for the Northern District of Ohio, the Western District of Pennsylvania, the District of Indiana, and the Northern District of Illinois, in causes in Chancery in the said Courts then depending, wherein Charles Moran and others were complainants, and the Pittsburgh, Fort Wayne and Chicago Rail Road Company and others were defendants, or acquired by virtue of a deed .bearing date the twenty-sixth day of February, 1862, made to the parties hereto of the first part by the Pittsburgh, Fort Wayne and Chicago Rail Road Company, pursuant to the aforesaid decrees, of, in and to all and singular the continuous railway, extending from its eastern terminus in Pittsburgh, in the State of Pennsylvania, to Chicago, in the State of Illinois, and commonly known as the Pittsburgh, Fort Wayne and Chicago Railroad or Railway; including all the railways, ways and rights of way, depot grounds and other lands; all tracks, bridges, viaducts, culverts, fences and other structures; all depots, station-houses, engine-houses, car-houses, freight-houses, wood-houses and other buildings; and all machine shops and other shops, held, or acquired for use in connection with the said railway or the business thereof; and including, also, all locomotives, tenders, cars, and other rolling stock or equipment, and all machinery, tools, implements, fuel and materials for the constructing, operating, repairing or replacing the said railway, or any part thereof, or any of its equipments or appurtenances; and also all franchises connected with or relating to .the said railway, or the construction, maintenance or use thereof; and all the property, franchises, rights and things, of whatever name or nature, which were conveyed by the aforesaid deeds of the said trustees and Master Commissioners, and of the said Company, to the parties of the first part hereto; *subject*, nevertheless, as to so much of the said railroad lying between the Federal street station, Alleghany City, in the State of Pennsylvania, and the passenger depot of the Pennsylvania Railroad on Liberty street, in the City of Pittsburgh, in said State, and of the other property connected therewith, as is embraced within a certain mortgage made by the Ohio and Pennsylvania Railroad Company to Thomas T. Firth, of the City of Philadelphia, and Reuben Miller, Jr., of the City of Pittsburgh, bearing date May 6th, 1856, to the lien created by the said mortgage; and *subject*, also, as to the rights of way, depot grounds, lots and premises in the city and vicinity of Chicago, purchased during the pendency of the said causes, in which Charles Moran and others are complainants, and the said Pittsburgh, Fort Wayne and Chicago Rail Road Company and others are defendants, as aforesaid, by the agency of the Receiver in the said causes, to the lien of the vendors for the purchase money of the same; and *subject*, also, as to any real estate included in the aforesaid sale by the said trustees and Master Commissioners to the lien, if any such exists, of any vendor or former owner of said real estate, not otherwise provided to be paid by the decrees or orders in the said causes: *Provided*, nevertheless, and it is the true intent and meaning of these presents, that nothing herein contained shall be construed to express or imply any covenant by the parties of the first part, or either of them, but that this instrument shall operate to convey, in behalf of the said parties, all the estates and interests in the railway and appurtenances, property, rights, franchises and things hereinbefore described, which the said parties, or either of them, might hold by virtue of the aforesaid conveyances, and which the said parties, each for himself, and not one for the other, can lawfully convey, and no more; and that the said estates and interests are hereby charged with, and shall pass by virtue of these presents, *subject* to the payment of all liabilities incurred in respect to the said railway, or its business, by the said parties of the first part, during their possession of the said railway: *Together* with all and singular the tenements, hereditaments and appurtenances thereunto belonging or in any wise appertaining, and the reversions, remainders, tolls, incomes, rents, issues and profits thereof; and also all the estate, right, title, interest, property, possession, claim, and demand, whatsoever, as well in law as in equity, of the said parties of the first part, of, in and to the same, and any and every part thereof, with the appurtenances: *To have and to hold* the above described premises and appurtenances, subject as aforesaid, unto the said parties of the second part as joint tenants, and not as tenants in common, and the survivor of them, and to the heirs and assigns of such survivor, to the only proper use and behoof of the said parties of the second part, and of the survivor of them, and of the heirs and assigns of such survivor; *in trust*, nevertheless, for the purposes herein expressed, to wit:

ARTICLE FIRST.—Until default shall be made in the payment of principal and interest of the said bonds, or some of them, or until default shall be made in respect to something herein required to be done or kept by the Pittsburgh, Fort Wayne and Chicago Railway Company, the said Pittsburgh, Fort Wayne and Chicago Railway Company shall be suffered and permitted to possess, manage, operate and enjoy the said railway from Pittsburgh to Chicago, with its equipments and appurtenances, and to take and use the rents, incomes, profits, tolls, and issues thereof, in the same manner and with the same effect as if this deed had not been made.

ARTICLE SECOND.—In case default shall be made in the payment of any interest on any of the aforesaid bonds, issued or to be issued, according to the tenor of the coupons thereto annexed, or in any payment required to be made into the special sinking fund herein provided, or in any requirement to be done or kept by the Pittsburgh, Fort Wayne and Chicago Railway Company; and if such default shall continue for the period of three months, it shall be lawful for the said trustees, or the survivor of them, or their or his successors, personally or by their or his attorneys or agents, to enter into and upon all and singular the premises hereby conveyed or intended so to be, and each and every part thereof; and to have, hold and use the same, operating by their or his superintendents, managers, receivers, or servants, or other attorneys or agents, the said railway, and conducting the business thereof, and making, from time to time, all repairs and replacements, and such useful alterations, additions and improvements thereto as may seem to them or him to be judicious; and to collect and receive all tolls, freights, incomes, rents, issues and profits of the same, and of every part thereof; and, after deducting the expenses of operating the said railway and conducting its business, and of all the said repairs, replacements, alterations, additions and improvements, and all payments which may be made for taxes, assessments, charges or liens prior to the lien of these presents upon the said premises, or any part thereof, as well as a just compensation for their or his own services, to apply the moneys arising as aforesaid, to the payment of interest, in the order in which such interest shall have become, or shall become due, ratably to the persons holding the coupons, evidencing the right to such interests and to the payment, in like order, of the corresponding contributions to the special sinking fund herein established for the six per cent. bonds; and after paying all interest which shall have become due, and all the said contributions to the special sinking fund, to apply the same to the satisfaction of the principal of the aforesaid bonds of the original and substituted issues which may be at that time unpaid, ratably and without-discrimination or preference.

ARTICLE THIRD.—In case default shall be made as aforesaid, and shall continue as aforesaid, it shall likewise be lawful for the said trustees, or the survivor of them, or their or his successors, after entry as aforesaid, or other entry, or without entry, personally, or by their or his attorneys or agents, to sell and dispose of all and singular the premises hereby conveyed, or in-

tended so to be, at public auction, in the City of New York, or
at such place within any of the States in which any part of the
said railway is situate, which the said trustees may designate,
and at such time as they may appoint, having first given notice
of the place and the time of such sale by advertisement, published
not less than three times a week for six weeks, in one or more
newspapers in the Cities of New York, Pittsburgh and Chicago,
and in Crestline and Fort Wayne, or some other places in the Judicial Districts of the United States in which Crestline and Fort
Wayne are situated, or to adjourn the said sale, from time to
time, in their or his discretion, and if so adjourning, to make the
same at the time and place to which the same may be so adjourned, and to make and deliver to the purchaser or purchasers
thereof good and sufficient deed or deeds in the law for the same
in fee simple ; which sale, made as aforesaid, shall be a perpetual
bar, both in law and equity, against the parties of the first part,
and all other persons lawfully claiming or to claim the said
premises, or any part thereof, by, from, through or under them,
or any or either of them ; and, after deducting from the proceeds
of such sale just allowances for all expenses of the said sale, including attorney's and counsel fees, and all other expenses,
advances, or liabilities, which may have been made or incurred by
the said trustees in operating or maintaining the said railway, or
in managing its business while in possession, and all payments
which may have been made by them for taxes or assessments, and
for charges and liens prior to the lien of these presents on the said
premises, or any part thereof, as well as compensation for their
own services, to apply the said proceeds to the payment of the
principal of such of the aforesaid bonds of the original issue, and
of the substituted issue, as may be at that time unpaid, whether
or not the same shall have previously become due, and of the
interest which shall at that time have accrued on the said principal and be unpaid, without discrimination or preference, but
ratably to the aggregate amount of such unpaid principal and
accrued and unpaid interest ; and if, after the satisfaction thereof, a surplus of the said proceeds shall remain, to apply the said
surplus according to the provisions of a deed of trust, bearing
even date herewith, and made by the parties of the first part to
the parties of the second part, subject to these presents, to secure
five millions and one hundred and sixty thousand dollars of
mortgage bonds, known as Second Mortgage Bonds of the said

Company, so far as the said trusts shall be then unfulfilled, or to pay over so much of said surplus as may be necessary for that purpose to the trustees then acting under the said deed ; and if, after so doing, a surplus of the said proceeds shall still remain, to apply or pay over the same, in like manner, upon the bonds, or to the trustees then acting under a deed of trust, bearing even date herewith, and made by the parties of the first part to the parties of the second part, subject to these presents, and to the said last mentioned trust deed, to secure two millions of dollars of bonds of the said Company, known as Third Mortgage or Income Bonds ; and if, after so doing, a surplus shall still remain, to pay over the same to the said Pittsburgh, Fort Wayne and Chicago Railway Company, or to such other parties as may be entitled to receive the same.

And it is hereby declared, that the receipt or receipts of the said trustees shall be a sufficient discharge to the purchaser or purchasers of the premises, for his or their purchase money, and that such purchaser or purchasers, his or their heirs, executors or administrators, shall not, after payment thereof, and having such receipt, be liable to see to its being applied upon or for the trusts and purposes of these presents, or in any manner, howsoever, be answerable for any loss, misapplication, or non-application of such purchase money, or any part thereof, or be obliged to inquire into the necessity, expediency, or authority, of or for any such sale.

ARTICLE FOURTH.—At any sale of the aforesaid property, or any part thereof, whether made by virtue of the power herein granted, or by judicial authority, the trustees shall bid for and purchase, or cause to be bid for and purchased, the property so sold, or any part thereof, in behalf of all the holders of the bonds secured by this instrument, and then outstanding, in the proportion of the respective interests of such bondholders,—at a reason-able price, if but a portion of the said property shall be sold ; or if all of it be sold, at a price not exceeding the whole amount of such bonds outstanding, with the interest accrued thereon.

ARTICLE FIFTH.—In case default shall be made in the payment of any half year's interest on any of the aforesaid bonds, whether of the original or substituted issues, at the time and in the manner in the coupon issued therewith provided, the said coupon having been presented, and the payment of the interest therein specified having been demanded, and that such default shall continue for the period of three months after the said coupon shall have become due; or in case default shall be made in the payment to the trustees of any part of any semi-annual instalment of the one per cent., per annum, for the special sinking fund, herein provided, for the redemption of the substituted issue of the six per cent. bonds, and such default shall continue for the period of three months after the said payment shall have become due, then, and in either of such cases, the principal of all the bonds secured hereby, whether of the original or substituted issues, shall, at the election of the trustees, become immediately due and payable, anything contained in the said bonds or herein to the contrary notwithstanding; but a majority in interest of the holders of the said bonds may, in writing, or by a vote of a meeting duly held, instruct the trustees to declare the said principal to be due, or to waive the right so to declare, on such terms and conditions as such majority shall deem proper, or may annul or reverse the election of the trustees, *provided*, that no action of the trustees or bondholders shall extend to or be taken to affect any subsequent default, or to impair the rights resulting therefrom

ARTICLE SIXTH.—In any six months,—the first such period commencing on the first day of January, 1862,—in which the net earnings of the said railway shall exceed the amount necessary to pay the interest upon all the bonds of the said Company, secured by these presents or by either of the two several trust deeds bearing even dates herewith, and then outstanding, including the contribution to the special sinking fund for the six per cent. bonds herein provided,—and a dividend of three per cent. upon six millions and five hundred thousand dollars of capital stock, such surplus shall be reserved, and shall, within sixty days after the expiration of the said six months, be paid over to the trustees as a sinking fund for the redemption of the bonds secured by these presents; *Provided*, that after the redemption of bonds amounting in the aggregate to two millions and five hundred thousand dollars, the said sinking fund may be limited to the application of not less than one per cent. per annum upon the aggregate of bonds outstanding at the time of such limitation; and the said sinking fund may at any time hereafter be varied in its amount, by agreement between the said Company and the holders of the bonds secured hereby, acting by a majority in interest; and, with the assent of the trustees, the said surplus or any part thereof may, as the same shall from time to time arise, be applied to the improvevent of the said railway, its equipments and appurtenances.

The trustees shall deposit the said surplus in some safe depository in the City of New York: and the said moneys, together with all accumulations of interest thereon that may actually come into the hands of the trustees, shall be invested by the trustees in the purchase of bonds secured by these presents, provided the same can be purchased at a rate not exceeding the par of said bonds, with the interest accrued thereon; and the bonds so purchased shall be held by the trustees, and immediately stamped or endorsed as belonging to the said sinking fund, but shall remain in force, and the interest thereon shall continue to be paid by the said Pittsburgh, Fort Wayne and Chicago Railway Company, and the amount thereof shall be added to and applied as a part of the capital of the sinking fund hereby established, and be invested in the manner and with the limitations herein provided.

And preparatory to such purchase of bonds the trustees shall give ten days notice thereof in one or more of the newspapers published in the City of New York, and shall make the said purchase at the lowest price or prices at which the bonds may be offered, pursuant to such notice, or at such lower price or prices as they may be able to obtain the same, but not exceeding the par and interest of said bonds; and in case the said bonds cannot be purchased at par and interest within three months after the expiration of the notice aforesaid, then the trustees shall invest the said moneys in the purchase of the bonds of the said Company secured by a second lien, at the lowest prices, not exceeding par and interest, at which they may be obtained after notice as aforesaid. And if bonds secured by the said second lien cannot be purchased as aforesaid, the trustees shall, in like manner, invest the said moneys in bonds of the said Company known as the Third Mortgage or Income Bonds. And in case of inability to invest in any of the three classes of securities, in the manner herein provided, the trustees shall invest the same in the said securities according to their best discretion.

ARTICLE SEVENTH.—The Pittsburgh, Fort Wayne and Chicago Railway Company shall, from time to time, prepare such bonds as may be necessary for the purposes of the exchange herein provided; which bonds shall in their form be identical with those originally issued in pursuance of these presents, except that their principal shall be irredeemable otherwise than by the operation of the special sinking fund herein provided—that they shall bear interest at the rate of six per centum per annum, and that they shall be entitled to the benefit of the said special sinking fund, in addition to that herein provided from the surplus earnings of the said railway; which special sinking fund shall consist of one per cent., per annum, upon the aggregate amount of six per cent. bonds issued in substitution for the seven per cent. bonds hereby secured, from the days when the last preceding coupon, at the rate of seven per cent. matured, which one per cent. shall be paid to the trustees in semi-annual instalments of one-half of one per cent. upon the days on which the semi-annual instalments of six per cent. interest shall be payable. The holder of any seven per cent. bond, of the original issue secured by these presents, shall be entitled, after the first day of July, 1867, to convert such bond into a six per cent bond of the issue herein provided, by notifying, in writing, the Company, on any day on which any semi-annual instalment of interest on the said bonds shall be payable, of his desire so to convert the said bond; and on depositing the said bond with the said Company for conversion, it shall be the duty of the Company, within a reasonable period, to duly execute a six per cent. bond for exchange, and of the trustees, for the time being, to countersign and deliver to the Company such six per cent. bond, upon the cancellation of the seven per cent. bond; which cancellation shall be made in such manner as the trustees may direct, and shall be recorded by the Company. Conversions may be made on such other notice as may be satisfactory to the Company and the trustees, but such conversions shall in all cases be as of the date of the maturity of the last preceding coupon of the bonds so converted. The moneys coming into this special sinking fund for the redemption of the six per cent. bonds shall be deposited in some safe depository in the City of New York; and as often as the moneys in the sinking fund, derived from the conversion of bonds made on or as of any particular date, shall amount to a sum sufficient to redeem one or more such bonds, the said moneys shall, without unreasonable delay, be applied to the redemption of bonds converted at or as of the said date, as follows:

The Trustees shall designate by lot so many of the bonds as

they have money to pay ; and shall give notice of the numbers designated, personally, to any owner or holder of the bonds known to them, and by advertisement in one or more daily newspapers in the City of New York, once a week, until the date of the next interest payment, if any owner or holder of the said bonds is unknown to them ; and on presentation and delivery of the said bonds shall apply the money so received by them to the payment thereof. And all future interest on any of the said bonds, not presented and delivered on or before the date of the said next interest payment, shall cease from and after such date ; and the Company shall be no longer liable for the said interest. The trustees shall, without unreasonable delay, cancel the bonds so redeemed by them, and return such bonds to the Company ; and the trustees and the Company shall keep separate registries of all the bonds so redeemed or so designated for redemption ; and the registry of the Company shall be at all reasonable times open to the inspection of each of the bondholders and stockholders of the Company, and the numbers and amounts of the bonds so redeemed or designated for redemption shall be reported by the Company in each annual statement made to the stockholders ; and the cancelled bonds shall, on the request of the meeting of the stockholders, be produced and exhibited.

ARTICLE EIGHTH.—The aggregate amount of the Capital Stock of the said Pittsburgh, Fort Wayne and Chicago Railway Company, outstanding at any one time, shall never exceed, in par value, six millions and five hundred thousand dollars, unless the holders of bonds secured by these presents shall have, by a vote at a meeting duly held, expressly consented to such increase.

ARTICLE NINTH.—The Pittsburgh, Fort Wayne and Chicago Railway Company shall, from time to time, and at all times hereafter, and as often as thereunto requested by the trustees, execute, deliver and acknowledge all such further deeds, conveyances and assurances in the law, for the better assuring unto the trustees and their successors in the trust hereby created, upon the trust herein expressed, the railway, equipments and appurtenances hereinbefore mentioned or intended so to be, and all other property and things, whatsoever, which may be hereafter acquired for use in connection with the same or any part thereof, and all franchises now held or hereafter acquired, including the franchise to be a corporation, as by the trustees, or the survivor of them, or their successors, or by their or his counsel learned in the law shall be reasonably advised, devised or required.

ARTICLE TENTH.—The trustees shall have full power, in their discretion, and upon the written request of the Pittsburgh, Fort Wayne and Chicago Railway Company, to convey by way of release, or otherwise, to the persons designated by the said Company the whole or any part of the lands situate in the City of Chicago and State of Illinois, and heretofore occupied by or purchased for the use of the Pittsburgh, Fort Wayne and Chicago Railroad Company; and, in order to substitute other lands, or to allow or aid the erection thereon of warehouses or other buildings to be used in connection with the business of the said railway, or to otherwise facilitate the said business, shall have like power to convey as aforesaid any lands acquired or held for the purposes of stations, depots, shops or other buildings, or the uses connected therewith; and shall also have power to convey as aforesaid any lands or property which, in the judgment of the trustees, shall not be necessary for use in connection with the said Pittsburgh, Fort Wayne and Chicago Railway, or which may have been held for a supply of fuel, gravel or other material ; and also to convey as aforesaid any lands which may become disused by reason of a change in the location of any station-house, depot, shop or other buildings connected with the said railway, and such lands occupied by the track and adjacent to such station-house, depot, shop or other building as the said Company may deem it expedient to disuse or abandon by reason of such change; and to consent to any such change and to such other changes in the location of the track as in their judgment shall have become expedient, and to make and deliver the conveyances necessary to carry the same into effect; but any lands which may be acquired for permanent use, in substitution for any so released, shall be conveyed to the trustees upon the trusts of these presents ; and the trustees shall also have full power to allow the said Company, from time to time, to dispose of, according to their discretion, such portions of the equipments, machinery and implements, at any time held or acquired for the use of the said railway as may have become unfit for such use, replacing the same by new, which shall be conveyed to the trustees, or be otherwise made subject to the operation of these presents.

ARTICLE ELEVENTH.—If the said Pittsburgh, Fort Wayne and Chicago Railway Company shall well and truly pay the sums of money herein required to be paid by the said Company, and all interest thereon, at the times and in the manner herein specified, and shall well and truly keep and perform all the things herein required to be kept or performed by the said Company, according to the true intent and meaning of these presents, then, and in that case, the estate, right, title and interest of the said parties of the second part, and of their successors in the trust hereby created, shall cease, determine, and become void, otherwise the same shall be and remain in full force and virtue.

ARTICLE TWELFTH.—The said Pittsburgh, Fort Wayne and Chicago Railway Company shall, at all times hereafter, keep a book at their office or agency in the City of New York, which shall be designated as the Voting Register of the First Mortgage Bondholders, and shall be distinct from the Transfer Register of the bonds.

Any holder of any of the said bonds shall be entitled to have his name and address, and the denomination and number of every of the said bonds held by him, entered in such registry, on presenting at the aforesaid office or agency, a written statement of the aforesaid particulars, signed by him, and, if required, duly verifying his title thereto by producing the bonds, or upon filing with the Company the written order of the person last registered as the holder, or in such other mode as may be prescribed by the regulations for such verification. The Trustees may, in the first instance, prescribe the said regulations, subject to the power hereby declared of the bondholders, acting by a majority in interest, to adopt, alter, or repeal, from time to time, the said regulations, and generally to establish such as may seem to them expedient. Such registration shall authenticate the right of the holder of every bond so registered to vote on the said bond, as provided therein, at every general and special meeting of the stockholders of the said Company ; and shall also entitle the said holder to notice, in such mode and form as may be fixed by regulations prescribed or established as aforesaid, of all meetings of the First Mortgage Bondholders. The trustees, and each of them, shall at all times have free access to such book of registry, and shall, from time to time, and at all times, on the request, in writing, of either of them be furnished with a copy thereof by the said Company; and shall have a right to require, at their option, that any act or reso lution of the said bondholders, affecting their duties or the interest of the trust hereby created, shall be authenticated by the signatures of all the persons assenting thereto, as well as by a minute of the proceedings of the meeting. Meetings of the First Mortgage Bondholders may be called by the trustees, or in such other mode as may be fixed by regulations prescribed or or established as aforesaid, and the bondholders may vote thereat in person or by proxy; and the quorum may be defined, and such other regulations or by-laws in respect to such meetings may be from time to time established, altered, or repealed by the bondholders, acting by a majority in interest, as to them shall seem expedient; and, until the bondholders shall act, such powers may be temporarily exercised by the trustees.

ARTICLE THIRTEENTH.—It is hereby declared and agreed that it shall be the duty of the Trustees to exercise the power of entry hereby granted, or the power of sale hereby granted, or both, or to take appropriate legal proceedings to enforce the rights of the bondholders under these presents, upon the requisition in writing, as hereinafter specified, as applicable to the several cases of default, in the manner and subject to the qualifications hereinafter provided, as follows:

1. If the default be as to interest or principal of any bonds, either of the original seven per cent. issue or of the substituted six per cent. issue provided for by these presents, or in respect to any installment of the special sinking fund hereby established for the redemption of the said six per cent. bonds, such requisition upon the said Trustees shall be by holders of not less than one hundred thousand dollars in aggregate amount of the said bonds, and upon such requisition, and a proper indemnification by the persons making the same, to the Trustees, against the costs and expenses to be by them incurred, it shall be the duty of the Trustees to enforce the rights of the bondholders under these presents, by entry, sale or legal proceedings, as they, being advised by counsel learned in the law, shall deem most expedient for the interest of all the holders of the said bonds.

2. If the default be in respect to any payment into the sinking fund established by Article Sixth of these presents, or be in the creation or issue of capital stock by the said Pittsburgh, Fort

Wayne and Chicago Railway Company in excess of the aggregate amount fixed by Article Eighth of these presents, or be in the omission of any act or thing required by Article Ninth of these presents for the further assuring of the title of the Trustees to any property or franchises now possessed or hereafter acquired, or in the omission to comply with each and all the provisions of Articles Seventh and Twelfth of these presents, or with any other provisions herein contained to be performed or kept by the said Company, then, and in either of such cases, the requisitions shall be as aforesaid ; but it shall be within the discretion of the Trustees to enforce or waive the rights of the, bondholders by reason of such default, subject to the power hereby declared of a majority in interest of the holders of the said bonds, by requisition in writing, or by a vote at a meeting duly held, to instruct the said Trustees to waive such default, or to enforce their rights by reason thereof ; provided that no action of the said Trustees or bondholders, or both, in waiving such default, or otherwise, shall extend to or be taken to affect any subsequent default, or to impair the rights resulting therefrom.

ARTICLE FOURTEENTH.—It is mutually agreed by and between the parties hereto, that the word " Trustees," as used in these presents, shall be construed to mean the Trustees for the time being, whether one or more be original or new ; and, whenever a vacancy shall exist, to mean the surviving or continuing Trustee ; and such Trustee shall, during such vacancy, be competent to exercise all the powors granted by these presents to the party of the second part. And it is mutually agreed by and between the parties hereto, as a condition on which the parties of the second part have assented to these presents, that the said Trustees shall not be in any manner responsible for any default or misconduct of each other ; that the said Trustess shall be entitled to just compensation for all services which they may hereafter render, in their trust, to be paid by the said Company ; that either of the said Trustees, or any successor, may resign, and discharge himself of the trust created by these presents, by notice in writing to the Pittsburgh, Fort Wayne and Chicago Railway Company, and to the existing Trustee, if there be such, three months before such resignation shall take effect, or such shorter time as they may accept as adequate notice, and upon the due execution of the conveyances hereinafter required ; that

the said Trustees, or either of them, may be removed by the vote
of a majority in interest of the holders of the aforesaid bonds,
the said vote being had at a meeting duly held of the said bondholders, and attested by an instrument under the hands and seals
of the persons so voting; that in case at any time hereafter
either of the said Trustees, or any Trustee hereafter appointed,
shall die or resign, or be removed, as herein provided, or by a
court of competent jurisdiction, or shall become incapable or
unfit to act in the said trust, a successor to such Trustee shall be
appointed by the surviving or continuing Trustee, with the consent of the holders for the time being of a majority in interest
of the said bonds, then outstanding, or the consent of a meeting,
duly held, of the holders of the said bonds; and the Trustee so
appointed, with the Trustee so surviving or continuing, shall
thereupon become vested with all the powers, authorities and
estates granted to or conferred upon the parties of the second
part by these presents, and all the rights and interests requisite
to enable him to execute the purposes of this trust, without any
further assurance or conveyance, so far as such effect may be
lawful; but the surviving or continuing Trustee shall immediately
execute all such conveyances and other instruments as may be
fit or expedient for the purpose of assuring the legal estate in
the premises, jointly with himself, to the Trustee so appointed;
and upon the death, resignation or removal of any Trustee, or
any appointment in his place in pursuance of these presents, all
his powers and authorities by virtue hereof shall cease; and all
the estate, right, title and interest in the said premises, of any
Trustee so dying, resigning or being removed, shall, if there be a
co-Trustee surviving or continuing in office, wholly cease and determine; but the said Trustee so resigning or being removed
shall, on the written request of the new Trustee who may be
appointed, immediately execute a deed or deeds of conveyance
to vest in such new Trustee, jointly with the continuing Trustee,
and upon the trusts herein expressed, all the property, rights and
franchises which may be at that time held upon the said trusts;
Proviaed, nevertheless, that it is hereby agreed and declared
that for the first vacancy in the trust hereby created which may
happen as hereinbefore provided, in respect to either of the parties of the second part hereto, Charles Lanier, of the City and
State of New York, is hereby designated and appointed to fill
such vacancy, with the same effect as if he had been appointed

in the manner hereinbefore specified ; and in case it shall at any time hereafter prove impracticable, after reasonable exertions, to appoint in the manner hereinbefore provided a successor in any vacancy which may have happened in said trust, or in case the trust shall become wholly vacant, application in behalf of all the holders of the bonds secured hereby, may be made by the surviving or continuing Trustee, or, if the trust be wholly vacant, by holders of the said bonds, to the aggregate amount of one hundred thousand dollars, to any Circuit Court of the United States, for any Judicial District in which any part of the aforesaid railway may be situate, for the appointment of a new Trustee or new Trustees.

In witness whereof, the parties of the first part have hereunto set their respective hands and seals, and the parties of the second part have also set their respective hands and seals, for the purpose or evidencing their acceptance of the trust hereby created, on the day and year first above written.

	J. F. D. Lanier, [L. S.]
	Samuel J. Tilden, [L. S.]
	Louis H. Meyer, [L. S.]
	J. Edgar Thomson, [S. S.]
	Samuel Hanna, [L. S.]
	John Ferguson, [L. S.]
Signed, sealed and delivered in presence of	Samuel J. Tilden. [L. S.]

James P. Sinnott and
John L. Drummond.

4.—THE SECOND MORTGAGE.

This Indenture, made the first day of March, in the year of our Lord one thousand eight hundred and sixty-two, between James F. D. Lanier, Samuel J. Tilden and Louis H. Meyer, of the City and State of New York, J. Edgar Thomson, of the City of Philadelphia, and State of Pennsylvania, and Samuel Hanna, of the town of Fort Wayne, and State of Indiana, of the first part, and John Ferguson and Samuel J. Tilden, of the City and State of New York, of the second part, witnesseth :

Whereas, The Pittsburgh, Fort Wayne and Chicago Railway Company is vested with franchises to be a corporation, granted to the said Company by the States of Pennsylvania, Illinois and Indiana, respectively, and has become duly organized as a corporation, in conformity to the provisions of the said grants, with capacity, in its corporate character, to take, hold, and exercise other franchises, and particularly with capacity to acquire, hold, maintain, and operate the continuous railway, extending from Pittsburgh, in the State of Pennsylvania, to Chicago, in the State of Illinois, commonly known as the PITTSBURGH, FORT WAYNE AND CHICAGO RAILROAD, together with its equipments and appurtenances.

And whereas, the said Company has agreed with the parties of the first part, to buy the aforesaid railway, and, in evidence of a portion of the consideration for the same, has made and delivered to the parties of the first part, its bonds, amounting in the aggregate to the sum of five millions and one hundred and sixty thousand dollars ; all of which bonds bear date on the first day of March, in the year one thousand eight hundred and sixty-two, and are payable at the office or agency of the said Company, in the City of New York, upon the first day of July, in the year one thousand nine hundred and twelve, and are redeemable, at the option of the Company, at any time after the first day of July, one thousand eight hundred and sixty-seven, on any day on which a half-yearly installment of interest shall fall due, and are convertible, at the option of the holders thereof, upon any such day, into bonds, to be issued and secured in the same manner as the said bonds, but bearing interest at the rate of six per cent., and irredeemable, except by a sinking fund of one per cent. per annum, on the whole amount of the said six per cent. bonds, which shall have been issued, to be reserved and applied in the manner hereinafter specified ; all of which bonds originally issued as aforesaid bear interest, after the first day of April, 1862, at the rate of seven per centum, per annum, payable semi-annually, at the office or agency of the said Company, in the City of New York ; of which bonds seven hundred and sixty, numbered, consecutively, from 1 to 760, inclusively, are each for one thousand dollars ; and two hundred, numbered, consecutively, from 761 to 960, inclusively, are each for five hundred dollars ; and forty-three hundred, numbered, consecutively, from 961 to 5,260, inclusively, are each for one thousand dollars ; and the interest on all of which said bonds is payable as follows, that is to say : The interest on nine hundred and sixty of the said bonds, numbered, consecutively, from 1 to 960, inclusively, is payable on the first days of January and July in each year; the interest on eight hundred and sixty of the said bonds, numbered, consecutively, from 961 to 1,820, inclusively, is payable on the first days of February and August in each year; the interest on eight hundred and sixty of the said bonds, numbered, consecutively, from 1,821 to 2,680, inclusively, is payable on the first days of March and September in each year; the interest on eight hundred and sixty of the said bonds, numbered, consecutively, from 2,681 to 3,540, inclusively, is payable on the first days of April and October in each year; the interest on eight hundred and sixty of the said bonds, numbered, consecutively, from 3,541 to 4,400, inclusively, is payable on the first days of May and November in each year; and the interest on eight hundred and sixty of the said bonds, numbered, consecutively, from 4,401 to 5,260, inclu sively, is payable on the first days of June and December in each year; all of which fifty-two hundred and sixty bonds are of like tenor, except as to the days on which the interest is payable, and except that the four hundred of the said bonds numbered, con secutively, from 1 to 400, inclusively, have, in case of a sale of the mortgaged property, a preference in the application of the pro ceeds thereof as hereinafter provided; and all of which fifty-two hundred and sixty bonds are in the form following:

No. ———. UNITED STATES OF AMERICA. $ ——.

"STATES OF PENNSYLVANIA, OHIO, INDIANA, AND ILLINOIS.

"PITTSBURGH, FORT WAYNE AND CHICAGO RAILWAY COMPANY.

"*Second Mortgage Bond.*

" Know all men by these presents, that the Pittsburgh, Fort Wayne and Chicago Railway Company are indebted to John Ferguson and Samuel J. Tilden, of the City of New York, or bearer, in the sum of   dollars, lawful money of the United States of America, which the said Company promise to pay to the said John Ferguson and Samuel J. Tilden, or to the bearer hereof, on the first day of July, in the year one thousand nine hundred and twelve, at the office or agency of the said Company, in the City of New York, with interest thereon, after the first day of April, 1862, at the rate of seven per centum per annum, payable, semi-annually, at the said office or agency in the City of New York, on the first days of   and in each year, on the presentation and surrender of the annexed coupons as they severally become due; and in case of the non-payment of any half-yearly installment of interest, which shall have become payable, and shall have been demanded, if such default shall continue for three months after the maturity of the said installment, the principal of this bond shall become due in the manner and with the effect provided in the deed of trust hereinafter mentioned. But it is hereby provided that, at any time after the first day of July, one thousand eight hundred and sixty-seven, the said Company, on any day on which a half-yearly installment of interest shall fall due, may at their option redeem at par the principal of this bond. And it is further agreed, that this bond is convertible, at the option of the holder, upon any day upon which any such installment of interest shall become payable, into a bond, to be issued and secured in the same manner as this bond, but bearing interest at the rate of six per cent., and irredeemable, except by a sinking fund of one per cent. per annum on the whole amount of the said six per cent. bonds which shall have been issued, to be reserved and applied in the manner specified in the said deed of trust.

" This bond is one of series   of an issue composed of series G (consisting of seven hundred and sixty bonds, for one thousand dollars each, and two hundred bonds for five hundred dollars each) ; and of series H, I, K, L and M (consisting severally of eight hundred and sixty bonds, for one thousand dollars each), which issue of five thousand two hundred and sixty bonds are numbered from one to five thousand two hundred and sixty, inclusively, and amount, in the aggregate, to five millions one hundred and sixty thousand dollars; all of which bonds bear date on the first day of March, 1862, and are of like tenor, except that the coupons of the six several series mature on the first days of the six successive months of each half year, and that the four hundred bonds of series G, numbered from one to four hundred, inclusively, have, in case of a sale of the mortgaged property, a preference in the application of the proceeds thereof ; the payment of all of which bonds is secured by a deed of trust, dated the first day of March, 1862, duly executed and delivered by James F. D. Lanier, Samuel J. Tilden, Louis H. Meyer, J. Edgar Thomson, and Samuel Hanna, to John Ferguson and Samuel J. Tilden, Trustees, and conveying the Pittsburgh, Fort Wayne and Chicago Railway, and the equipments, appurtenances and things therein described, subject to the prior lien thereon created by a deed of trust between the same parties, and bearing the same date, given to secure $5,250,000 of First Mortgage Bonds of the said Company.

" And this bond is also entitled to the benefits of the sinking fund by the said deed of trust provided.

" The person appearing on the voting bond-register of the said Company as the holder of this bond, at the time of any meeting of the stockholders of the said Company, will be entitled to one vote at such meeting for every two hundred dollars of the par amount hereof. The right to vote upon this bond is transferable on the written order of the person last registered as its holder, or on the production of the bond by the holder.

" It is agreed between the said Company and the holder of this bond, that no recourse shall be had for its payment to the individual liability of any stockholder of the said Company ; and that, in case of any default in the payment hereof, the said Company hereby, waives the benefit of any extension, stay, or appraisement laws, now existing or that may hereafter exist.

" This bond shall pass by delivery or by transfer on the books of the Company in the City of New York. After a registration of ownership, certified hereon by the transfer agent of the Company, no transfer except on the books of the Company shall be valid, unless the last transfer be to bearer, which shall restore transferability by delivery. But this bond shall continue subject to successive registrations and transfers to bearer, as aforesaid, at the option of each holder.

" This bond shall not become obligatory until it shall have been authenticated by a certificate endorsed hereon and duly signed by the Trustees.

> In Witness Whereof, the said Company have caused their corporate seal to be hereunto affixed, and the same to be attested by the signatures of their President and Secretary, and have also caused the coupons hereto annexed to be signed by their Secretary, on this first day of March, in the year one thousand eight hundred and sixty-two.
>
> —————————,
> *President.*

—————————,
Secretary.

And whereas, provision is herein made for the issue of bonds bearing interest at the rate of six per centum per annum, but irredeemable, except by a special sinking fund herein established, into which the bonds hereinbefore described may be converted, at the option of the respective holders thereof, in the manner and on the conditions herein prescribed.

And whereas, the intention of these presents is, and is hereby declared to be, that all of the said bonds, whether of the original seven per cent. issue or of the subsequent six per cent. issue, for the purpose of conversion as aforesaid, shall be equally secured by these presents, in proportion to the amount of the principal thereof unpaid, with the interest on the said principal accrued and unpaid, without discrimination or preference, with reference to the times of the actual issue of the said bonds, or of the maturing of the principal thereof, or the maturing of any interest which shall have accrued thereon, or other preference, excepting only the preference in the application of the proceeds of sale of the mortgaged property, hereinafter provided, in favor of the four hundred bonds numbered from one to four hundred, inclusively.

Now THIS INDENTURE WITNESSETH, That the parties of the first part, in consideration of the premises, and of one dollar to them in hand paid, the receipt whereof is hereby acknowledged, and in order to secure the payment of the principal and interest of the bonds aforesaid, issued or to be issued, as herein recited and provided, and every part of the said principal and interest, as the same shall become payable, according to the tenor of the said bonds and of the coupons thereto annexed, have granted, bargained and sold, and do, by these presents grant, bargain, sell, convey and transfer, unto the parties of the second part, all the right, title and interest of them, the parties of the first part, and of any or either of the said parties, acquired by virtue of a deed, bearing date the eighteenth day of February, 1862, and made to the parties hereto of the first part by John Ferguson and Thomas E. Walker, both of the City and State of New York, Trustees and Special Master Commissioners, in pursuance of decrees of the Circuit Courts of the United States for the Northern District of Ohio, the Western District of Pennsylvania, the District of Indiana, and the Northern District of Illinois, in causes in Chancery in the said Courts then depending, wherein Charles Moran and others were complainants, and the Pittsburgh, Fort Wayne and Chicago Railroad Company and others were defendants, or acquired, by virtue of a deed bearing date the twenty-sixth day of Februaay, 1862, made to the parties hereto of the first part, by the Pittsburgh, Fort Wayne and Chicago Railroad Company, pursuant to the aforesaid decrees of, in, and to ALL AND SINGULAR the continuous railway extending from its eastern terminus, in Pittsburgh, in the State of Pennsylvania, to Chicago, in the State of Illinois, and commonly known as the PITTSBURGH, FORT WAYNE AND CHICAGO RAILROAD OR RAILWAY, including all the railways, ways and rights of way, depot grounds and other lands; all tracks, bridges, viaducts, culverts, fences and other structures; all depots, station-houses, engine-houses, car-houses, freight-houses, wood-houses and other buildings; and all machine shops and other shops, held or acquired for use in connection with the said railway, or the business thereof; and including also all locomotives, tenders, cars and other rolling stock or equipment; and all machinery, tools, implements, fuel and materials for the constructing, operating, repairing or replacing the said railway, or any part thereof, or any of its equipments or appurtenances; and also all franchises connected with or relating to the said railway, or the construction, maintenance or use thereof; and all the property, franchises, rights and things, of whatever name or nature, which were conveyed by the aforesaid deeds of the said Trustees and Master Commissioners, and of the said Company, to the parties of the first part hereto; *subject, nevertheless,* to a certain deed of trust or mortgage, bearing even date herewith, made by the parties of the first part to the parties of the second part hereto, for the purpose of securing the payment of bonds of the Pittsburgh, Fort Wayne and Chicago Railway Company, known as First Mortgage Bonds, amounting, in the aggregate, to five millions and two hundred and fifty thousand dollars; and also *subject* as to so much of the said railroad lying between the Federal street Station, in Allegheny City, in the State of Pennsylvania, and the passenger depot of the Pennsylvania Railroad, on Liberty street, in Pittsburgh City, in said State, and of the other property connected therewith, as is embraced within a certain mortgage made by the Ohio and Pennsylvania Railroad Company to Thomas T. Firth, of the City of Philadelphia, and Reuben Miller, Jr., of the City of Pittsburgh, bearing date May 6th, 1856, to the lien created by the said mortgage; and *subject also* as to the rights of way, depot grounds, lots and premises in the City and vicinity of Chicago, purchased during the pendency of the said causes, in which Charles Moran and others are complainants, and the said Pittsburgh, Fort Wayne and Chicago Railroad Company and others are defendants, as aforesaid, by the agency of the Receiver in the said causes, to the lien of the vendors for the purchase money of the same ; and *subject also* as to any real estate included in the aforesaid sale by the said Trustees and Master Commissioners to the lien, if any such exists, of any vendor or former owner of said real estate, not otherwise provided to be paid by the decrees or orders in the said causes ; *Provided, nevertheless,* and it is the true intent and meaning of these presents, that nothing herein contained shall be construed to express or imply any covenant by the parties of the first part, or either of them, but that this instrument shall operate to convey, in behalf of the said parties, all the estates and interests in the railway and appurtenances, property, rights, franchises, and things hereinbefore described, which the said parties, or either of them, might hold by virtue of the aforesaid conveyances, and which the said parties, each for himself, and not one for the other, can lawfully convey, and no more ; and that the said estates and interests are hereby charged with and shall pass, by virtue of these presents, *subject* to the payment of all liabilities incurred in respect to the said railway, or its business, by the said parties of the first part, during their possession of the said railway : *Together* with all and singular the tenements, hereditaments, and appurtenances thereunto belonging, or in anywise appertaining, and the reversions, remainders, tolls, incomes, rents, issues, and profits thereof ; and also all the estate, right, title, interest, property, possession, claim and demand whatsoever, as well in law as in equity, of the said parties of the first part of, in, and to the same, and any and every part thereof, with the appurtenances ; *To have and to hold* the above described premises and appurtenances, subject as aforesaid, unto the said parties of the second part, as joint tenants, and not as tenants in common, and the survivor of them, and to the heirs and assigns of such survivor, to the only proper use and behoof of the said parties of the second part, and of the survivor of them, and of the heirs and assigns of such survivor ; *in trust,* nevertheless, for the purposes herein expressed, to wit :

ARTICLE FIRST.—Until default shall be made in the payment of principal or interest of the said bonds, or some of them, or until default shall be made in respect to something herein required to be done or kept by the Pittsburgh, Fort Wayne and

Chicago Railway Company, the said Pittsburgh, Fort Wayne and Chicago Railway Company shall be suffered 'and permitted, to possess, manage, operate, and enjoy the said railway from Pittsburgh to Chicago, with its equipments and appurtenances, and to take and use the rents, incomes, profits, tolls, and issues thereof, in the same manner and with the same effect as if this deed had not been made.

ARTICLE SECOND.—In case default shall be made in the payment of any interest on any of the aforesaid bonds issued or to be issued, according to the tenor of the coupons thereto annexed, or in any payment required to be made into the special sinking fund herein provided, or in any requirement to be done or kept.by the Pittsburgh, Fort Wayne and Chicago Railway Company, and if such default shall continue for the period of three months, it shall be lawful for the said Trustees, or the survivor of them, or their or his successors, personally or by their or his attorneys or agents, to enter into and upon all and singular the premises hereby conveyed, or intended so to be, and each and every part thereof; and to have, hold and use the same, operating, by their or his superintendents, managers, receivers, or servants, or other attorneys or agents, the said railway, and conducting the business thereof, and making from time to time all repairs and replacements, and such useful alterations, additions and improvements thereto as may seem to them or him to be judicious; and to collect and receive all tolls, freights, incomes, rents, issues and profits of the same, and of every part thereof; and after deducting the expenses of operating the said railway and conducting its business, and of all the said repairs, replacements, alterations, additions, and improvements, and all payments which may be made for taxes, assessments, charges, or liens prior to the lien of these presents upon the said premises, or any part thereof, as well as a just compensation for their or his own services, to apply the moneys arising as aforesaid to the payment of interest, in the order in which such interest shall have become or shall become due, ratably to the persons holding the coupons evidencing the right to such interest, and to the payment, in like order, of the corresponding contributions to the special sinking fund herein established for the six per cent. bonds; and, after paying all interest which shall have become due, and all the said contributions

to the special sinking fund, to apply the same to the satisfaction
of the principal of the aforesaid bonds of the original and substituted issues, which may be at that time unpaid, ratably and
without discrimination or preference.

ARTICLE THIRD.—In case default shall be made, as aforesaid,
and shall continue as aforesaid, it shall likewise be lawful for
the said Trustees, or the survivor of them, or their or his successors, after entry as aforesaid or other entry, or without entry,
personally or by their or his attorneys or agents, to sell and dispose of all and singular the premises hereby conveyed, or intended so to be, at public auction, in the City of New York, or at
such place, within any of the States in which any part of the said
railway is situate, which the said Trustees may designate, and at
such time as they may appoint, having first given notice of the
place and the time of such sale, by advertisement, published not
less than three times a week for six weeks, in one or more newspapers in the cities of New York, Pittsburgh, Chicago, and in
Crestline and Fort Wayne, or some other places in the judicial
districts of the United States in which Crestline and Fort Wayne
are situated ; or to adjourn the said sale, from time to time, in
their or his discretion, and, if so adjourning, to make the same at
the time and place to which the same may be so adjourned, and
to make and deliver to the purchaser or purchasers thereof good
and sufficient deed or deeds in the law for the same in fee simple ;
which sale, made as aforesaid, shall be a perpetual bar, both in
law and equity, against the parties of the first part, and all other
persons lawfully claiming or to claim the said premises, or any
part thereof, by, from, through or under them, or any or either
of them ; and after deducting from the proceeds of such sale just
allowances for all expenses of the said sale, including attorneys'
and counsel fees, and all other expenses, advances, or liabilities
which may have been made or incurred by the said Trustees in
operating or maintaining the said railway, or in managing its
business while in possession, and all payments which may have
been made by them for taxes or assessments, and for charges and
liens prior to the lien of these presents on the said premises, or
any part thereof, as well as compensation for their own services,
to apply the proceeds in the manner following:

First.—To the payment of the principal of the aforesaid
bonds of the original seven per cent. issue, numbered from one

to four hundred, inclusively, and such bonds of the six per cent. issue as may have been substituted for bonds numbered as aforesaid, as may be at that time unpaid, whether or not the same shall have previously become due, without discrimination or preference, but ratably to the aggregate amount thereof.

Secondly.—After paying the principal of such of the aforesaid four hundred bonds as may be at that time unpaid, to the payment of the principal of such of the aforesaid bonds, numbered from four hundred and one to fifty-two hundred and sixty, inclusively, of the original issue and of the substituted issue, whether or not the same shall have previously become due, and of the interest which shall at that time have accrued on the principal of all the bonds secured hereby and be unpaid, without discrimination or preference, but ratably to the aggregate amount of such unpaid principal and accrued and unpaid interest.

And if, after satisfaction thereof, a surplus of the said proceeds shall remain, to apply such surplus according to the provisions of a deed of trust, bearing even date herewith, and made by the parties of the first part to the parties of the second part hereto, subject to these presents, and to the lien prior to these presents, to secure two millions of dollars of mortgage bonds known as income or third mortgage bonds of the said Company, so far as the trusts thereof shall be then unfulfilled, or to pay over so much of such surplus as may be necessary for that purpose to the Trustees then acting under the said deed; and if, after so doing, a surplus shall still remain, to pay over the same to the said Pittsburgh, Fort Wayne and Chicago Railway Company, or to such other parties as may be entitled to receive the same.

And it is hereby declared, that the receipt or receipts of the said Trustees shall be a sufficient discharge to the purchaser or purchasers of the premises, for his or their purchase money, and that such purchaser or purchasers, his or their heirs, executors, or administrators, shall not, after payment thereof, and having such receipt, be liable to see to its being applied upon or for the trusts and purposes of these presents, or in any manner, whatsoever, be answerable for any loss, misapplication, or non-application of such purchase money, or any part thereof, or be obliged to inquire into the necessity, expediency, or authority of or for any such sale.

ARTICLE FOURTH.—At any sale of the aforesaid property, or any part thereof, whether made by virtue of the power herein granted, or by judicial authority, the Trustees shall bid for and purchase, or cause to be bid for and purchased, the property so sold, or any part thereof, in behalf of all the holders of the bonds secured by this instrument and then outstanding, in the proportion of the respective interests of such bondholders, at a reasonable price, if but a portion of the said property shall be sold ; or if all of it be sold, at a price not exceeding the whole amount of such bonds outstanding, with the interest accrued thereon.

ARTICLE FIFTH.—In case default shall be made in the payment of any half year's interest on any of the aforesaid bonds, whether of the original or substituted issues, at the time and in the manner in the coupon issued therewith provided, the said coupon having been presented. and the payment of the interest therein specified having been demanded, and that such default shall continue for the period of three months after the said coupon shall have become due, or in case default shall be made in the payment to the Trustees of any part of any semi-annual installment of the one per cent. per annum for the special sinking fund herein provided for the redemption of the substituted issue of six per cent. bonds, and such default shall continue for the period of three months after the said payment shall have become due, then, and in either of such cases, the principal of all the bonds secured hereby, whether of the original or substituted issues, shall, at the election of the Trustees, become immediately due and payable, anything contained in the said bonds or herein to the contrary notwithstanding ; but a majority in interest of the holders of the said bonds may, in writing, or by a vote of a meeting duly held, instruct the Trustees to declare the said principal to be due, or to waive the right so to declare, on such terms and conditions as such majority shall deem proper, or may annul or reverse the election of the Trustees; *Provided*, that no action of the Trustees or bondholders shall extend to or be taken to affect any subsequent default, or to impair the rights resulting therefrom.

ARTICLE SIXTH.—In any six months, the first such period commencing on the first day of January, 1862, in which the net earnings of the said railway shall exceed the amount necessary to pay the interest upon all the bonds of the said Company, secured by these presents, or by either of the two several trust deeds bearing even dates herewith, and then outstanding, including the contribution to the special sinking fund for the six per cent. bonds herein, and in the said first mortgage, provided, and a dividend of three per cent. upon six millions and five hundred thousand dollars of the capital stock, such surplus shall be reserved, and shall, within sixty days after the expiration of the said six months, be paid over to the Trustees under the trust deed known as the first mortgage, as a sinking fund for the redemption of the bonds secured by the said trust deed, and by these presents, and by a trust deed herein mentioned, creating a subsequent lien, to be applied in the manner in the said first mortgage provided ; *Provided*, that after the redemption of bonds amounting in the aggregate to two million and five hundred thousand dollars, the said sinking fund may be limited to the application of not less than one per cent. per annum upon the aggregate of bonds outstanding at the time of such limitation ; and the said sinking fund may at any time hereafter be varied in its amount ; and the said surplus, or any part thereof, may, as the same shall from time to time arise, be applied to the improvement of the said railway, its equipments and appurtenances, in the manner in the said first mortgage provided.

Article Seventh.—The Pittsburgh, Fort Wayne and Chicago Railway Company shall, from time to time, prepare such bonds as may be necessary for the purposes of the exchange herein provided, which bonds shall, in their form, be identical with those originally issued in pursuance of these presents, except that their principal shall be irredeemable otherwise than by the operation of the special sinking fund herein provided—that they shall bear interest at the rate of six per centum per annum, and that they shall be entitled to the benefit of the said special sinking fund, in addition to that herein provided, from the surplus earnings of the said railway ; which special sinking fund shall consist of one per cent. per annum upon the aggregate amount of six per cent. bonds issued in substitution for the seven per cent. bonds hereby secured, from the days when the last preceding coupon, at the rate of seven per cent., matured, which one per cent. shall be paid to the Trustees in semi-annual installments of one-half of one per cent., upon the days on which the semi-annual installments of six per cent. interest shall be payable. The holder of any seven per cent. bond of the original issue secured by these presents shall be entitled, after the first day of July, 1867, to convert such bond into a six per cent. bond of the issue herein provided, by notifying in writing the Company, on any day on which any semi-annual installment of interest on the said bonds shall be payable, of his desire so to convert the said bond; and on depositing the said bond with the said Company for conversion, it shall be the duty of the Company, within a reasonable period, to duly execute a six per cent. bond for exchange, and of the Trustees for the time being to countersign and deliver to the Company such six per cent. bond, upon the cancellation of the seven per cent. bond; which cancellation shall be made in such manner as the Trustees may direct, and shall be recorded by the Company. Conversions may be made on such other notice as may be satisfactory to the Company and the Trustees; but such conversions shall, in all cases, be as of the date of the maturity of the last preceding coupon of the bonds so converted. The moneys coming into this special sinking fund for the redemption of the six per cent. bonds shall be deposited in some safe depository in the City of New York, and as often as the moneys in the sinking fund derived from the conversion of bonds made on or as of any particular date shall amount to a sum sufficient to redeem one or more such bonds, the said moneys shall, without unreasonable delay, be applied to the redemption of bonds converted at or as of the said date, as follows:

The Trustees shall designate by lot so many of the bonds as they may have money to pay, and shall give notice of the numbers designated, personally, to any owner or holder of the bonds known to them, and by advertisement in one or more daily newspapers in the City of New York, once a week, until the date of the next interest payment, if any owner or holder of the said bonds is unknown to them; and, on presentation and delivery of the said bonds, shall apply the money so received by them to the payment thereof. And all future interest on any of the said bonds, not presented and delivered on or before the date of the said next interest payment, shall cease from and after such date, and the Company shall be no longer liable for the said interest. The Trustees shall, without unreasonable delay, cancel the bonds so redeemed by them, and return such bonds to the Company; and the Trustees and the Company shall keep separate registries of all the bonds so redeemed or so designated for redemption; and the registry of the Company shall be, at all reasonable times, open to the inspection of each of the bondholders and stockholders of the Company, and the numbers and amounts of the bonds so redeemed or designated for redemption shall be reported by the Company in each annual statement made to the stockholders, and the canceled bonds shall, on the request of the meeting of the stockholders, be produced and exhibited.

ARTICLE EIGHTH.—The aggregate amount of the capital stock of the said Pittsburgh, Fort Wayne and Chicago Railway Company, outstanding at any one time, shall never exceed, in par value, six millions and five hundred thousand dollars, unless the holders of bonds secured by these presents shall have, by a vote at a meeting duly held, expressly consented to such increase.

ARTICLE NINTH.—The Pittsburgh, Fort Wayne and Chicago Railway Company shall, from time to time, and at all times hereafter, and, as often as thereunto requested by the Trustees, execute, deliver, and acknowledge all such further deeds, conveyances, and assurances in the law, for the better assuring unto the Trustees and their successors in the trust hereby created, upon the trusts herein expressed, the railway, equipments, and appurtenances hereinbefore mentioned, or intended so to be, and all other property and things whatsoever, which may be hereafter acquired for use in connection with the same, or any part thereof, and all franchises now held or hereafter acquired, including the franchise to be a corporation, as by the Trustees, or the survivor of them, or their successors, or by their or his counsel learned in the law, shall be reasonably advised, devised or required.

ARTICLE TENTH.—The Trustees shall have full power, in their discretion, and upon the written request of the Pittsburgh, Fort Wayne and Chicago Railway Company, to convey, by way of release or otherwise, to the persons designated by the said Company, the whole or any part of the lands situate in the City of Chicago, and State of Illinois, and heretofore occupied by or purchased for the use of the Pittsburgh, Fort Wayne and Chicago Railroad Company; and, in order to substitute other lands, or to allow or aid the erection thereon of warehouses or other buildings, to be used in connection with the business of the said railway, or to otherwise facilitate the said business, shall have like power to convey, as aforesaid, any lands acquired or held for the purposes of stations, depots, shops, or other buildings, or the uses connected therewith; and shall also have power to convey, as aforesaid, any lands or property which, in the judgment of the Trustees, shall not be necessary for use, in connection with the said Pittsburgh, Fort Wayne and Chicago Railway, or which may have been held for a supply of fuel, gravel, or other material; and also to convey, as aforesaid, any lands which may become disused by reason of a change of the location of any station house, depot, shop, or other building connected with the said railway, and such lands occupied by the track, and adjacent to such station house, depot, shop, or other building, as the said Company may deem it expedient to disuse or abandon by reason of such change; and to consent to any such change, and to such other changes in the location of the track as in their judgment shall have become expedient, and to make and deliver the conveyances necessary to carry the same into effect; but any lands which may be acquired for permanent use in substitution for any so released shall be conveyed to the Trustees upon the trusts of these presents; and the said Trustees shall also have full power to allow the said Company, from time to time, to dispose of, according to their discretion, such portions of the equipments, machinery and implements, at any time held or acquired for the use of the said railway, as may have become unfit for such use, replacing the same by new, which shall be conveyed to the Trustees, or be otherwise made subject to the operation of these presents.

Article Eleventh.—If the said Pittsburgh, Fort Wayne and Chicago Railway Company shall well and truly pay the sums of money herein required to be paid by the said Company, and all interest thereon, at the times and in the manner herein specified, and shall well and truly keep and perform all the things herein required to be kept and performed by the said Company, according to the true intent and meaning of these presents, then, and in that case, the estate, right, title and interest of the said parties of the second part, and of their successors in the trust hereby created, shall cease, determine, and become void ; otherwise the same shall be and remain in full force and virtue.

Article Twelfth.—The said Pittsburgh, Fort Wayne and Chicago Railway Company shall, at all times hereafter, keep a book at their office or agency in the City of New York, which shall be designated as the Voting Register of the Second Mortgage Bondholders; and shall be distinct from the Transfer Register of the bonds. Any holder of any of the said bonds shall be entitled to have his name and address, and the denomination and the number of every of the said bonds held by him, entered in such registry, on presenting at the aforesaid office or agency a written statement of the said particulars, signed by himself, and, if required, duly verifying his title thereto by producing the bonds, or upon filing with the Company the written order of the person last registered as the holder; or in such other mode as may be prescribed by the regulations for such verification. The Trustees may, in the first instance, prescribe the said regulations, subject to the power hereby declared of the bondholders, acting by a majority in interest, to adopt, alter, or repeal, from time to time, the said regulations, and generally to establish such as may seem to them expedient. Such registration shall authenticate the right of the holder of every bond so registered, to vote on the said bond, as provided therein, at every general and special meeting of the stockholders of the said Company; and shall also entitle the said holder to notice, in such mode and form as may be fixed by regulations prescribed or established as aforesaid, of all meetings of the Second Mortgage Bondholders. The Trustees, and each of them, shall, at all times, have free access to such book of registry, and shall, from time to time, and at all times, on the request in writing of either of them, be furnished with a copy thereof by the said Company; and shall have a right to require, at their option, that any act or resolution of the said bondholders, affecting their duties or the interest of the trust hereby created, shall be authenticated by the signatures of all the persons assenting thereto, as well as by a minute of the proceedings of the meeting. Meetings of the Second Mortgage Bondholders may be called by the Trustees, or in such other mode as may be fixed by regulations prescribed or established as aforesaid, and the bondholders may vote thereat in person or by proxy; and the quorum may be defined, and such other regulations for by-laws, in respect to such meetings, may be from time to time established, altered, or repealed by the bondholders, acting by a majority in interest, as to them shall seem expedient; and, until the bondholders shall act, such powers may be temporarily exercised by the Trustees.

114

Article Thirteenth.—It is hereby declared and agreed, that it shall be the duty of the Trustees to exercise the power of entry hereby granted, or the power of sale hereby granted, or both, or to take appropriate legal proceedings to enforce the rights of the bondholders under these presents, upon the requisition in writing, as hereinafter specified as applicable to the several cases of default, in the manner and subject to the qualifications hereinafter provided, as follows :

1. If the default be as to interest or principal of any bonds, either of the original seven per cent. issue or of the substituted six per cent. issue, provided for by these presents, or in respect to any installment of the special sinking fund hereby established for the redemption of the said six per cent. bonds, such requisition upon the said Trustees shall be by holders of not less than one hundred thousand dollars in aggregate amount of the said bonds; and upon such requisition and a proper indemnification by the persons making the same to the Trustees against the costs and expenses to be by them incurred, it shall be the duty of the Trustees to enforce the rights of the bondholders under these presents, by entry, sale or legal proceedings, as they, being advised by counsel learned in the law, shall deem most expedient for the interest of all the holders of the said bonds.

2. If the default be in respect to any payment into the sinking fund established by Article Sixth of these presents, or be in the creation or issue of capital stock by the said Pittsburgh, Fort Wayne and Chicago Railway Company, in excess of the aggregate amount fixed by Article Eighth of these presents, or be in the omission of any act or thing, required by Article Ninth of these presents, for the further assuring of the title of the Trustees to any property or franchises, now possessed or hereafter acquired, or in the omission to comply with each and all the provisions of Articles Seventh and Twelfth of these presents, or with any other provision herein contained, to be performed or kept by the said Company, then, and in either of such cases, the requisitions shall be as aforesaid ; but it shall be within the discretion of the Trustees to enforce or waive the rights of the bondholders by reason of such default, subject to the power hereby declared of a majority in interest of the holders of the said bonds, by requisition in writing, or by a vote at a meeting duly held, to instruct the said Trustees to waive such default or to enforce their rights by reason thereof ; provided that no action of the said Trustees, or bondholders, or

both, in waiving such default, or otherwise, shall extend to or be taken to affect any subsequent default, or to impair the rights resulting therefrom.

ARTICLE FOURTEENTH.—It is mutually agreed by and between the parties hereto, that the word "Trustees," as used in these presents, shall be construed to mean the Trustees for the time being, whether one or both be original or new; and, whenever a vacancy shall exist, to mean the surviving or continuing Trustee; and such Trustee shall, during such vacancy, be competent to exercise all the powers granted by these presents to the parties of the second part. And it is mutually agreed by and between the parties hereto, as a condition on which the parties of the second part have assented to these presents, that the said Trustees shall not be in any manner responsible for any default or misconduct of each other; that the said Trustees shall be entitled to just compensation for all services which they may hereafter render in their trust, to be paid by the said Company; that either of the said Trustees, or any successor, may resign and discharge himself of the trust created by these presents, by notice in writing to the Pittsburgh, Fort Wayne and and Chicago Railway Company, and to the existing Trustee, if there be such, three months before such resignation shall take effect, or such shorter time as they may accept as adequate notice, and upon the due execution of the conveyances hereinafter required; that the said Trustees, or either of them, may be removed by the vote of a majority in interest of the holders of the aforesaid bonds, the said vote being had at a meeting duly held of the said bondholders, and attested by an instrument under the hands and seals of the persons so voting; that in case at any time hereafter either of the said Trustees, or any Trustee hereafter appointed, shall die, or resign, or be removed as herein provided, or by a court of competent jurisdiction, or shall become incapable or unfit to act in the said trust, a successor to such Trustee shall be appointed by the surviving or continuing Trustee with the consent of the holders, for the time being, of a majority in interest of the said bonds then outstanding, or the consent of a meeting, duly held, of the holders of the said bonds; and the Trustee so appointed, with the Trustee so surviving or continuing, shall thereupon become vested with all the powers, authorities and estates granted to or conferred upon the parties of the second part by these presents, and all the rights and interests requisite to enable him to execute the purposes of this trust, without any further assurance or conveyance, so far as such effect may be lawful ; but the surviving or continuing Trustee shall immediately execute all such conveyances and other instruments as may be fit or expedient for the purpose of assuring the legal estate in the premises, jointly with himself, to the Trustee so appointed ; and upon the death, resignation or removal of any Trustee, or any appointment in his place in pursuance of these presents, all his powers and authorities by virtue hereof shall cease ; and all the estate, right, title and interest in the said premises of any Trustee so dying, resigning or being removed, shall, if there be a co-trustee surviving or continuing in office, wholly cease and determine ; but the said Trustee so resigning or being removed, shall, on the written request of the new Trustee who may be appointed, immediately execute a deed or deeds of conveyance to vest in such new Trustee, jointly with the continuing Trustee, and upon the trusts herein expressed, all the property, rights and franchises which may be at that time held upon the said trusts ; *provided*, nevertheless, that it is hereby declared and agreed that for the first vacancy in the trust hereby created, which may happen as hereinbefore provided, in respect to either of the parties of the second part hereto, Charles Lanier, of the City and State of New York, is hereby designated and appointed to fill such vacancy, with the same effect as if he had been appointed in the manner hereinbefore specified ; and in case it shall at any time hereafter prove impracticable, after reasonable exertions, to appoint in the manner hereinbefore provided, a successor in any vacancy which may have happened in said trust, or in case the trust shall become wholly vacant, application, in behalf of all the holders of the bonds secured hereby, may be made by the surviving or continuing Trustee, or, if the trust be wholly vacant, by holders of the said bonds to the aggregate amount of one hundred thousand dollars, to any Circuit Court of the United States for any Judicial District in which any part of the aforesaid railway may be situate, for the appointment of a new Trustee or new Trustees.

IN WITNESS WHEREOF, the parties of the first part have hereunto set their respective hands and seals, and the parties of the second part have also set their respective hands and seals, for the purpose of evidencing their acceptance of the trust hereby created, on the day and year first above written.

J. F. D. LANIER. [L. S.]
SAMUEL J. TILDEN. [L. S.]
LOUIS H. MEYER. [L. S.]
J. EDGAR THOMSON. [L. S.]
SAMUEL HANNA. [L. S.]
JOHN FERGUSON, [L. S.]
SAMUEL J. TILDEN, [L. S.]

Signed, sealed and delivered in }
 presence of }

JAMES P. SINNOTT and
JOHN L. DRUMMOND.

4.—THE THIRD MORTGAGE.

THIS INDENTURE, made this first day of March, in the year of our Lord one thousand eight hundred and sixty-two, between JAMES F. D. LANIER, SAMUEL J. TILDEN and LOUIS H. MEYER, of the City and State of New York ; J. EDGAR THOMSON, of the City of Philadelphia, and State of Pennsylvania ; and SAMUEL HANNA, of the Town of Fort Wayne, and State of Indiana, of the first part, and JOHN FERGUSON and SAMUEL J. TILDEN, of the City and State of New York, of the second part, WITNESSETH :

Whereas, The Pittsburgh, Fort Wayne and Chicago Railway Company is vested with franchises to be a corporation, granted to the said Company by the States of Pennsylvania, Illinois, and Indiana, respectively, and has become duly organized as a corporation in conformity to the provisions of the said grants, with capacity, in its corporate character, to take, hold and exercise other franchises, and particularly with capacity to acquire, hold, maintain and operate the continuous railway, extending from Pittsburgh, in the State of Pennsylvania, to Chicago, in the State of Illinois, commonly known as the Pittsburgh, Fort Wayne and Chicago Railroad, together with its equipments and appurtenances.

And whereas, the said Company has agreed with the parties of the first part to buy the aforesaid railway, and, in evidence of a portion of the consideration for the same, has made and delivered to the parties of the first part its bonds, amounting in the aggregate to the sum of two millions of dollars, all of which bonds bear date on the first day of March, in the year one thousand eight hundred and sixty-two, and are payable at the office or agency of the said Company, in the City of New York, at the pleasure of the said Company, after the first day of July, in the year one thousand nine hundred and twelve; and all of which bonds bear interest at such rate, not exceeding seven per centum per annum, as the net earnings hereinafter defined of the said railway, in each calendar year, may suffice to pay, after satisfying the interest on the First and Second Mortgage Bonds of the said Company; such interest for each calendar year upon the said bonds being payable at the office or agency of the said Company, in the City of New York, on the first day of April, after the termination of such year; of which said bonds fifteen hundred, numbered, consecutively, from 1 to 1,500, inclusively, are each for one thousand dollars, and one thousand, numbered, consecutively, from 1,501 to 2,500, inclusively, are each for five hundred dollars; and all of which bonds are in the form following:

No.   UNITED STATES OF AMERICA. $

STATES OF PENNSYLVANIA, OHIO, INDIANA, AND ILLINOIS.

PITTSBURGH, FORT WAYNE AND CHICAGO RAILWAY COMPANY.

INCOME BOND.

Know all men by these presents, that the Pittsburgh, Fort Wayne and Chicago Railway Company will pay to John Ferguson and Samuel J. Tilden, or assigns,   dollars, lawful money of the United States of America, at the pleasure of the said Company, after the first day of July, 1912, at the office or agency of said Company, in the City of New York, with interest thereon, after the first day of April, 1862, at such rate, not exceeding seven per cent. per annum, as the net earnings of the said railway in each calendar year, may suffice to pay on the entire issue of which this bond is a part, after satisfying the interest upon the first and second mortgage bonds of the said Company, such interest on this bond for each year ending on each 31st day of December, being payable at the office or agency of the said Company in the City of New York, on the first day of April, after the termination of the said year.

" This bond is one of an issue composed of a series of fifteen hundred bonds for one thousand dollars each, numbered from 1 to 1,500, inclusively, and a series of one thousand bonds for five hundred dollars each, numbered from 1,501 to 2,500, inclusively, and amounting in the aggregate to two millions of dollars, all bearing date on the first day of March, 1862, and the payment of which is secured by a deed of trust, dated the first day of March, 1862, duly executed and delivered by James F. D. Lanier, Samuel J. Tilden, Louis H. Meyer, J. Edgar Thomson and Samuel Hanna, to John Ferguson and Samuel J. Tilden, Trustees, and conveying the Pittsburgh, Fort Wayne and Chicago Railway, and the equipments, appurtenances and things therein described, subject to the prior liens thereon created by two deeds of trust between the same parties, and bearing the same date— one given to secure $5,250,000 First Mortgage Bonds of the said Company, and the other given to secure $5,160,000 of Second Mortgage Bonds of said Company.

" This bond is also entitled to the benefits of the sinking fund by the said deed of trust provided.

" The person appearing on the Register of the said Company as the holder of this bond, at the time of any meeting of the stockholders of the said Company, will be entitled to one vote, in person or by proxy, at such meeting, for every one hundred dollars of the par amount thereof.

" It is agreed between the said Company and the holder of this bond, that no recourse shall be had for its payment to the individual liability of any stockholder of the said Company, or to any liability of the said Company, except for the application of the net earnings of the said railway, as defined and provided in the deed of trust by which this bond is secured ; and that in case of any default in respect to the payment hereof the said Company hereby waives the benefit of any extension, stay or appraisement laws, now existing or that may hereafter exist. This bond shall pass by transfer on the books of the Company in the City of New York, authenticated by the transfer agent of the Company.

"This bond shall not become obligatory until it shall have been authenticated by a certificate, endorsed hereon, and duly signed by the Trustees.

> "IN WITNESS WHEREOF, the said Company have caused their corporate seal to be hereto affixed, and the same to be attested by the signatures of their President and Secretary, on this the first day of March, in the year one thousand eight hundred and sixty-two.

————————————————,
President.

————————————————,
Secretary.

And whereas, the intention of these presents is, and is hereby declared to be, that all of the said bonds shall be equally secured by these presents, in proportion to the amount of the principal thereof unpaid, with the interest on the said principal accrued and unpaid, without discrimination or preference, with respect to the times of the actual issue of the said bonds, or the maturing of any interest which shall have accrued thereon.

Now THIS INDENTURE WITNESSETH, That the parties of the first part, in consideration of the premises, and of one dollar to them in hand paid, the receipt whereof is hereby acknowledged, and in order to secure the payment of the principal and interest of the bonds aforesaid, issued or to be issued, as herein recited and provided, and every part of the said principal and interest, as the same shall become payable, according to the tenor of the said bonds, have granted, bargained and sold, and do by these presents grant, bargain, sell, convey and transfer, unto the parties of the second part, all the right, title and interest of them, the parties of the first part, and of any or either of the said parties, acquired by virtue of a deed, bearing date the eighteenth day of February, 1862, and made to the parties hereto of the first part, by John Ferguson and Thomas E. Walker, both of the City and State of New York, Trustees and Special Master Commissioners, in pursuance of decrees of the Circuit Courts of the United States for the Northern District of Ohio, the Western District of Pennsylvania, the District of Indiana, and the Northern District of Illinois, in causes in Chancery, in the said Courts then depend-

ing, wherein Charles Moran and others were complainants, and
the Pittsburgh, Fort Wayne and Chicago Railroad Company and
others were defendants, or acquired by virtue of a deed, bearing
date the twenty-sixth day of February, 1862, made to the parties hereto of the first part, by the Pittsburgh, Fort Wayne and
Chicago Railroad Company, pursuant to the aforesaid decrees,
of, in, and to all and singular, the continuous railway, extending
from its eastern terminus, in Pittsburgh, in the State of Pennsylvania, to Chicago, in the State of Illinois, and commonly
known as the PITTSBURGH, FORT WAYNE AND CHICAGO RAILROAD, OR
RAILWAY, including all the railways, ways, and rights of way,
depot grounds, and other lands, all tracks, bridges,
viaducts, culverts, fences, and other structures, all depots, station-houses, engine-houses, car-houses, freight-
houses, wood-houses, and other buildings; and all machine
shops and other shops, held or acquired for use in connection with the said railway or the business thereof, and including
also all locomotives, tenders, cars, and other rolling stock, or
equipment, and all machinery, tools, implements, fuel, and materials for the constructing, operating, repairing, or replacing the
said railway, or any part thereof, or any of its equipments or
appurtenances; and also all franchises, connected with or relating to the said railway, or the construction, maintenance or use
thereof; and also all the property, franchises, rights, and things,
of whatever name or nature, which were conveyed by the aforesaid deeds of the said Trustees and Master Commissioners, and
of the said Company, to the parties of the first part hereto;
subject, nevertheless, to a certain deed of trust or mortgage,
bearing even date herewith, made by the parties of the first part
to the parties of the second part hereto, for the purpose of securing the payment of bonds of the Pittsburgh, Fort Wayne and
Chicago Railway Company, known as First Mortgage Bonds,
amounting in the aggregate to five millions and two hundred and
fifty thousand dollars, and *subject, also*, to a certain other deed
of trust or mortgage, bearing even date herewith, made by the
said parties of the first part to the said parties of the second part,
for the purpose of securing the payment of bonds of the said
Pittsburgh, Fort Wayne and Chicago Railway Company, known
as Second Mortgage Bonds, amounting in the aggregate to five
millions and one hundred and sixty thousand dollars; and *subject, also*, as to so much of the said railroad, lying between the

Federal Street Station, in Alleghany City, in the State of Pennsylvania, and the passenger depot of the Pennsylvania Railroad on Liberty street, in Pittsburgh City, in said State, and of the other property connected therewith, as is embraced within a certain mortgage made by the Ohio and Pennsylvania Railroad Company to Thomas T. Firth, of the City of Philadelphia, and Reuben Miller, Jr., of the City of Pittsburgh, bearing date May 6th, 1856, to the lien created by the said mortgage; and *subject, also*, as to the rights of way, depot grounds, lots and premises in the City and vicinity of Chicago, purchased during the pendency of the said causes, in which Charles Moran and others are complainants, and the said Pittsburgh, Fort Wayne and Chicago Railroad Company and others are defendants, as aforesaid, by the agency of the Receiver in the said causes, to the lien of the vendors for the purchase money of the same; and *subject, also*, as to any real estate included in the aforesaid sale by the said Trustees and Master Commissioners, to the lien, if any such exists, of any vendor or former owner of said real estate, not otherwise provided to be paid by the decrees or orders in the said causes: *Provided, nevertheless,* and it is the true intent and meaning of these presents, that nothing herein contained shall be construed to express or imply any covenant by the parties of the first part, or either of them, but that this instrument shall operate to convey, in behalf of the said parties, all the estates and interests in the railway and appurtenances, property, rights, franchises, and things hereinbefore described, which the said parties, or either of them, might hold by virtue of the aforesaid conveyances, and which the said parties, each for himself, and not one for the other, can lawfully convey, and no more; and that the said estates and interests are hereby charged with, and shall pass, by virtue of these presents, *subject* to the payment of all liabilities incurred in respect to the said railway, or its business, by the said parties of the first part during their possession of the said railway: *Together* with all and singular the tenements, hereditaments and appurtenances thereunto belonging, or in anywise appertaining, and the reversions, remainders, tolls, incomes, rents, issues and profits thereof; and also all the estate, right, title, interest, property, possession, claim and demand whatsoever, as well in law as in equity, of the said parties of the first part, of, in, and to the same, and any and every part thereof, with the appurtenances. *To have and to hold* the above-described premises and

123

appurtenances, subject as aforesaid unto the said parties of the
second part, as joint tenants, and not as tenants in common, and
the survivor of them, and to the heirs and assigns of such survivor,. to the only proper use and behoof of the said parties of
the second part, and of the survivor of them, and of the heirs
and assigns of such survivors, *in trust*, nevertheless, for the
purposes herein expressed, to wit:

ARTICLE FIRST.—Until default shall be made in respect to
something herein required to be done or kept by the Pittsburgh,
Fort Wayne and Chicago Railway Company, the said Pittsburgh,
Fort Wayne and Chicago Railway Company shall be suffered and
permitted to possess, manage, operate, and enjoy the said railway from Pittsburgh to Chicago, with its equipments and appurtenances, and to take and use the rents, incomes, profits, tolls, and
issues thereof, in the same manner, and with the same effect, as
if this deed had not been made.

ARTICLE SECOND.—In case default shall be made in the application to the payment of interest upon the aforesaid bonds, at the
time and in the manner in Article Ninth of these presents prescribed, after satisfying prior liens as therein provided, of the
net earnings of the said railway, for any calendar year, as such
net earnings are defined in the said Article Ninth ; and if such default shall continue for the period of twelve months, then, and
in that case, upon the requisition of the bondholders secured
hereby, specified in the said Article Ninth, it shall be lawful for the
said Trustees, or the survivor of them, or their and his successors,
personally or by their or his attorneys or agents, to enter into and
upon, all and singular the premises hereby conveyed, or intended
so to be, and each and every part thereof ; and to have, hold and use
the same, operating by their or his superintendents, managers, receivers, or servants, or other attorneys or agents, the said railway,
and conducting the business thereof, and making, from time to
time, all repairs and replacements, and such useful alterations,
additions and improvements thereto, as may seem to them or him
to be judicious ; and to collect and receive all tolls, freights, incomes, rents, issues and profits of the. same, and of every part
thereof ; and after deducting the expenses of operating the said
railway and conducting its business, and of all the said repairs,
replacements, alterations, additions, and improvements, and all

payments which may be made for taxes, assessments, charges, or liens prior to the lien of these presents, upon the said premises, or any part thereof, as well as just compensation for their or his own services, to apply the moneys arising as aforesaid to the payment of interest, in the order in which such interest shall have become or shall become due, ratably to the persons holding the bonds entitled to such interest; and, after paying all interest which shall then be payable as herein above provided, to apply the same to the satisfaction of the principal of the aforesaid bonds which may be at that time unpaid, ratably and without discrimination or preference.

ARTICLE THIRD.—In case default shall be made, as aforesaid, and shall continue as aforesaid, then and in that case, upon requisition as aforesaid, it shall likewise be lawful for the said Trustees, or the survivor of them, or their or his successors, after entry as aforesaid or other entry, or without entry, personally or by their or his attorneys or agents, to sell and dispose of all and singular the premises hereby conveyed, or intended so to be, at public auction in the City of New York, or at such place, within any of the States in which any part of the said railway is situate, which the said Trustees may designate, and at such time as they may appoint, having first given notice of the place and the time of such sale, by advertisement, published not less than three times a week for six weeks, in one or more newspapers in the City of New York, Pittsburgh, Chicago, and in Crestline and Fort Wayne, or some other places in the judicial districts of the United States in which Crestline and Fort Wayne are situated; or to adjourn the said sale, from time to time, in their or his discretion, and, if so adjourning, to make the same at the time and place to which the same may be so adjourned; and to make and deliver, to the purchaser or purchasers thereof, good and sufficient deed or deeds in the law for the same in fee simple; which sale, made as aforesaid, shall be a perpetual bar both in law and equity against the parties of the first part, and all other persons lawfully claiming or to claim the said premises, or any part thereof, by, from, through, or under them, or any or either of them; and after deducting from the proceeds of such sale just allowances for all expenses of the said sale, including attorney's and counsel fees, and all other expenses, advances, or liabilities, which may have been made or incurred by the said Trustees in operating or maintaining the said railway, or in managing its business while in possession, and all payments which may have been made by them for taxes or assessments, and for charges and liens prior to the lien of these presents, on the said premises, or any part thereof, as well as compensation for their own services, to apply the said proceeds to the payment of the principal of such of the aforesaid bonds as may be at that time unpaid, and of the interest which shall at that time have accrued on the said principal as herein provided, and be unpaid, without discrimination or preference, but ratably to the aggregate amount of such unpaid principal, and accrued and unpaid interest; and if after the satisfaction thereof, a surplus of the said proceeds shall remain, to pay over the same to the said Pittsburgh, Fort Wayne and Chicago Railway Company, or to such other parties as may be entitled to receive the same.

And it is hereby declared that the receipt or receipts of the said Trustees shall be a sufficient discharge to the purchaser or purchasers of the premises, for his or their purchase money, and that such purchaser or purchasers, his or their heirs, executors or administrators, shall not, after payment thereof and having such receipt, be liable to see to its being applied upon or for the trusts and purposes of these presents, or in any manner howsoever be answerable for any loss, misapplication, or non-application of such purchase money, or any part thereof, or be obliged to inquire into the necessity, expediency, or authority of or for any such sale.

ARTICLE FOURTH.—At any sale of the aforesaid property, or any part thereof, whether made by virtue of the power herein granted or by judicial authority, the Trustees shall bid for and purchase, or cause to be bidden for and purchased, the property so sold, or any part thereof, in behalf of all the holders of the bonds secured by this instrument and then outstanding, in the proportion of the respective interests of such bondholders, at a reasonable price, if but a portion of the said property shall be sold, or if all of it be sold, at a price not exceeding the whole amount of such bonds outstanding, with the interest accrued thereon.

ARTICLE FIFTH.—The Pittsburgh, Fort Wayne and Chicago Railway Company shall, from time to time, and at all times hereafter, and as often as thereunto requested by the Trustees, execute, deliver, and acknowledge all such further deeds, conveyances, and assurances in the law, for the better assuring unto the Trustees, and their successors in the trust hereby created, upon the trusts herein expressed, the railway, equipments, and appurtenances hereinbefore mentioned or intended so to be, and all other property and things whatsoever, which may be hereafter acquired, for use in connection with the same or any part thereof, and all franchises now held or hereafter acquired, including the franchise to be a corporation, as by the Trustees, or the survivor of them, or their successors, or by their or his counsel learned in the law, shall be reasonably advised, devised, or required.

ARTICLE SIXTH.—Every conveyance, by way of release or otherwise, which may be executed and delivered in pursuance of Article Tenth of the deed of trust, prior in lien to these presents, by which the First Mortgage Bonds of the Pittsburgh, Fort Wayne and Chicago Railway Company are secured, shall operate to discharge the lands or other property so conveyed from the lien of these presents, as fully and effectually as such conveyance shall operate to discharge the same from the lien of the said trust deed; and every power or authority which may be exercised, and every act or thing which may be done or suffered by the Trustees of the said trust deed, in pursuance of the said Article Tenth, shall have the same effect with respect to the rights of the holders of bonds secured by these presents, as with respect to the rights of holders of bonds secured by the said deed of trust.

ARTICLE SEVENTH.—If the said Pittsburgh, Fort Wayne and Chicago Railway Company shall well and truly pay the sums of money herein required to be paid by the said Company, and all interest thereon, at the times and in the manner herein specified, and shall well and truly keep and perform all the things herein required to be kept or performed by the said Company, according to the true intent and meaning of these presents, then, and in that case, the estate, right, title and interest of the said parties of the second part, and of their successors in the trust hereby created, shall cease, determine and become void; otherwise the same shall be and remain in full force and virtue.

ARTICLE EIGHTH.—The said Pittsburgh, Fort Wayne and Chicago Railway Company shall, at all times hereafter, keep, at their office or agency in the City of New York, a book for the transfer of its bonds ; and shall also keep a transfer book at Pittsburgh, and such other places as the said Company may appoint. And the said Company shall likewise keep a book at their office or agency in the City of New York, which shall be designated as the Voting Register of the Third Mortgage Bondholders, and shall be dis- tinct from the Transfer Register of the bonds.

Any holder of any of the said bonds shall be entitled to have his name and address, and the denomination, and the number of every of the said bonds held by him entered in such registry, on presenting at the aforesaid office or agency a written statement of the said particulars, signed by himself, and, if required, duly verifying his title thereto by producing the bonds, or upon filing with the Company the written order of the person last registered as the holder; or in such other mode as may be prescribed by the regulations for such verification. The Trustees may, in the first instance, prescribe the said regulations, subject to the power hereby declared of the bondholders, acting by a majority in in- terest, to adopt, alter, or repeal, from time to time, the said regu- lations, and generally to establish such as may seem to them ex- pedient. Such registration shall authenticate the right of the holder of every bond so registered, to vote on the said bond as provided therein, at every general and special meeting of the stockholders of the said Company ; and shall also entitle the said holder to notice, in such mode and form as may be fixed by regulation prescribed or established as aforesaid of all meetings of the Third Mortgage Bondholders. The Trustees, and each of them, shall at all times have free access to such book of registry, and shall, from time to time, and at all times, on the request in writing of either of them, be furnished with a copy thereof by the said Company; and shall have a right to require, at their option, that an act or resolution of the said bondholders, affecting their duties, or the interest of the trust hereby created, shall be authenticated by the signatures of all the persons assenting thereto, as well as by a minute of the proceedings of the meeting. Meetings of the Third Mortgage Bondholders may be called by the Trustees, or in such other mode as may be fixed by regula- tions prescribed or established as aforesaid, and the bondholders may vote thereat in person or by proxy ; and the quorum may be defined, and such other regulations or by-laws, in respect to such meetings may be, from time to time, established, altered, or repealed by the bondholders, acting by a majority in interest, as to them shall seem expedient ; and until the bondholders shall act, such powers may be temporarily exercised by the Trustees.

ARTICLE NINTH.—It is hereby declared and agreed that the words "net earnings," as used in these presents, shall be construed to mean such surplus of the earnings of the said railway as shall remain after paying all expenses of operating the said railway and carrying on its business, including all taxes and assessments and payments on incumbrances prior in lien to these presents upon specific portions of the property hereby conveyed, of completing, repairing, or replacing the said railway, its appurtenances and equipments, so that the same shall be in high condition, and of providing. such additional equipment as the said Company shall deem necessary for the business of the said railway.

And it is further declared and agreed that the default contemplated in Articles Second and Third of these presents, shall consist in the omission to apply the aforesaid net earnings for any calendar year, after satisfying the payments for interest and special sinking fund of the First Mortgage Bonds, and of the Second Mortgage Bonds, secured by trust deeds prior in lien to these presents, and hereinbefore mentioned, to the payment of interest upon the income bonds secured by these presents; which application shall be made upon the first day of April next after the termination of such calendar year: *Provided*, that such portion of the said earnings as shall amount to less than one-half of one per cent. upon the aggregate amount of the income bonds then outstanding shall be reserved and carried to the interest fund for the next year for the said income bonds.

And it is further agreed and declared, that the requisition which shall be necessary to authorize or require the Trustees to exercise the power of entry or sale hereinbefore granted, shall be in writing, and signed by the holders of a majority in interest of the income bonds then outstanding, secured by these presents.

ARTICLE TENTH.—It is mutually agreed, by and between the parties hereto, that the word " Trustees," as used in these presents, shall be construed to mean the Trustees for the time being, whether one or both be original or new ; and, whenever a vacancy shall exist, to mean the surviving or continuing Trustee; and such Trustee shall, during such vacancy,be competent to exercise all the powers granted by these presents to the parties of the second part. And it is mutually agreed by and between the parties hereto, as a condition on which the parties of the second part have assented to these presents, that the said Trustees shall not be in any manner responsible for any default or misconduct of each other ; that the said Trustees shall be entitled to just compensation for all services which they may hereafter render in their trust, to be paid by the said Company ; that either of the said Trustees, or any successor, may resign and discharge himself of the trust created by these presents, by notice in writing to the Pittsburgh, Fort Wayne and Chicago Railway Company, and to the existing Trustee, if there be such, three months before such resignation shall take effect, or such shorter time as they may accept as adequate notice, and upon the due execution of the conveyances hereinafter required; that the said Trustees, or either of them, may be removed by the vote of a majority in interest of the holders of the aforesaid bonds, the said vote being had at a meeting, duly held of the said bondholders, and attested by an instrument under the hands and seals of the persons so voting ; that in case, at any time hereafter, either of the said Trustees, or any Trustee hereafter appointed, shall die, or resign, or be removed as herein provided, or by a court of competent jurisdiction, or shall become incapable or unfit to act in the said trust, a successor to such Trustee shall be appointed by the surviving or continuing Trustee, with the consent of the holders for the time being of a majority in interest of the said bonds then outstanding, or the consent of a meeting, duly held, of the holders of the said bonds; and the Trustee so appointed, with the Trustee so surviving or continuing, shall thereupon become vested with all the powers, authorities and estates granted to or conferred upon the parties of the second part by these presents, and all the rights and interests requisite to enable him to execute the purposes of this trust, without any further assurance or conveyance, so far as such effect may be lawful; but the surviving or continuing Trustee shall immediately execute all such conveyances and other instruments as may be fit or expedient for the purpose of assuring the legal estate in the premises, jointly with himself, to the Trustee so appointed; and upon the death, resignation, or removal of any Trustee, or any appointment in his place in pursuance of these presents, all his powers and authorities, by virtue hereof, shall cease; and all the estate, right, title and interest in the said premises of any Trustee so dying, resigning, or being removed, shall, if there be a co-Trustee surviving or continuing in office, wholly cease and determine; but the said Trustee so resigning or being removed, shall, on the written request of the new Trustee who may be appointed, immediately execute a deed or deeds of conveyance to vest in such new Trustee, jointly with the continuing Trustee, and upon the trusts herein expressed, all the property, rights and franchises which may be at that time held upon the said trusts. *Provided*, nevertheless, that it is hereby agreed and declared, that for the first vacancy in the trust hereby created, which may happen as hereinbefore provided, in respect to either of the parties of the second part hereto, Charles Lanier, of the City and State of New York, is hereby designated and appointed to fill such vacancy, with the same effect as if he had been appointed in the manner hereinbefore specified; and in case it shall at any time hereafter prove impracticable, after reasonable exertions, to appoint in the manner hereinbefore provided a successor in any vacancy which may have happened in said trust, or in case the trust shall become wholly vacant, application in behalf of all the holders of the bonds secured hereby may be made by the surviving or continuing Trustee; or, if the trust be wholly vacant, by holders of the said bonds to the aggregate amount of one hundred thousand dollars, to any Circuit Court of the United States, for any Judicial District in which any part of the aforesaid railway may be situate, for the appointment of a new Trustee or new Trustees.

In WITNESS WHEREOF, the parties of the first part have hereunto set their respective hands and seals, and the parties of the second part have also set their respective hands and seals, for the purpose of evidencing their acceptance of the trust hereby created, on the day and year first above written.

J. F. D. LANIER,	[L. S.]
SAMUEL J. TILDEN,	[L. S.]
LOUIS H. MEYER,	[L. S.]
J. EDGAR THOMSON,	[L. S.]
SAMUEL HANNA,	[L. S.]
JOHN FERGUSON,	[L. S.]
SAMUEL J. TILDEN.	[L. S.]

Signed, sealed and delivered in }
  presence of

JAMES P. SINNOTT and
JOHN L. DRUMMOND.

5.—FINAL CONVEYANCE TO RAILWAY COMPANY, SUBJECT TO THE MORTGAGES.

THIS INDENTURE, made this second day of March, in the year of our Lord one thousand eight hundred and sixty-two, between JAMES F. D. LANIER, and MARY M. LANIER, his wife, SAMUEL J. TILDEN, bachelor, and LOUIS H. MEYER, and ANN CHARLOTTE MEYER, his wife, of the City and State of New York; J. EDGAR THOMSON, and LAVINIA F. THOMSON, his wife, of the City of Philadelphia, and State of Pennsylvania; and SAMUEL HANNA, and ELIZA HANNA, his wife, of the town of FORT WAYNE, and State of Indiana, of the first part, and the PITTSBURGH, FORT WAYNE AND CHICAGO RAILWAY COMPANY of the second part, WITNESSETH:

Whereas, the said Pittsburgh, Fort Wayne and Chicago Railway Company, party of the second part, is vested with franchises to be a corporation, granted to the said Company, by the States of Pennsylvania, Illinois and Indiana, respectively; and has become duly organized as a corporation, in conformity to the provision of the said grants, with capacity in its corporate character, to take, hold and exercise other franchises, and particularly with capacity to acquire, hold, maintain and operate the continuous railway, extending from Pittsburgh, in the State of Pennsylvania, to Chicago, in the State of Illinois, commonly known as the Pittsburgh, Fort Wayne and Chicago Railroad, together with its equipments and appurtenances.

And whereas, the said Company has agreed with the parties of the first part, to buy the aforesaid railway, and, in evidence of a portion of the consideration for the same, has made and delivered to the parties of the first part its bonds, bearing date on the first day of March, in the year one thousand eight hundred and sixty-two, and payable at the office or agency of the said Company, in the City of New York; of which bonds, six thousand and six hundred, secured by a first lien, created by a deed of trust or mortgage hereinafter mentioned, amount in the aggregate to the sum of five million two hundred and fifty thousand dollars; are payable on the first day of July, in the year one thousand nine hundred and twelve, and are redeemable, at the option of the Company, at any time after the first day of July, one thousand eight hundred and sixty-seven, on any day on which a half-yearly instalment of interest shall fall due, and are convertible, at the option of the holders thereof, upon any such day, into bonds, to be issued and secured in the same manner as the said bonds, but bearing interest at the rate of six per cent., and irredeemable, except by a sinking fund of one per cent., per annum, on the whole amount of the said six per cent. bonds which shall have been issued, to be reserved and applied in the manner in said deed of trust specified; all of which bonds originally issued, as aforesaid, bear interest at the rate of seven per cent. per annum, payable semi-annually, at the office or agency of said Company, in the City of New York, on the several days in the said bonds mentioned; five thousand two hundred and sixty bonds, secured by a second lien, created by a deed of trust or mortgage hereinafter mentioned, amount in the aggregate to the sum of five million one hundred and sixty thousand dollars; are payable upon the first day of July, in the year one thousand nine hundred and twelve, and are redeemable, at the option of the Company, at any time after the first day of July, one thousand eight hundred and sixty-seven, on any day on which a half-yearly instalment of interest shall fall due, and are convertible, at the option of the holders thereof, upon any such day, into bonds to be issued and secured in the same manner as the said bonds, but bearing interest at the rate of six per cent., and irredeemable, except by a sinking fund of one per cent. per annum on the whole amount of the said six per cent. bonds which shall have been issued, to be reserved and applied in the manner in said second deed of trust or mortgage specified ; all of which bonds, originally issued, as aforesaid, bear interest after the first day of April, 1862, at the rate of seven per centum per annum, payable semi-annually, at the office or agency of the said Company in the City of New York, on the several days in the said bonds mentioned ; twenty-five hundred bonds, secured by the third lien, created by a deed of trust or mortgage hereinafter mentioned, amount in the aggregate to the sum of two millions of dollars, are payable at the pleasure of said Company, after the first day of July, in the year one thousand nine hundred and twelve, and bear interest at such rate, not exceeding seven per centum per annum, as the net earnings, in said third deed of trust or mortgage defined, of the said railway, in each calendar year, may suffice to pay, after satisfying the interest on the First and Second Mortgage Bonds of the said Company—such interest for each calendar year, upon the said bonds, being payable at the office or agency of the said Company in the City of New York, on the first day of April, after the termination of each year.

And whereas, the said Company, for the residue of the said consideration, has issued and delivered to the said parties of the first part, or to persons designated by them, sixty-five thousand shares of its capital stock, amounting in the aggregate to six million five hundred thousand dollars.

And whereas, the parties of the first part to these presents, for the consideration in the said several deeds of trust or mortgage expressed, and in order to secure the payment of the principal and interest of the bonds aforesaid, issued or to be issued, as in said several deeds of trust or mortgage recited and provided, and every part of the said principal and interest, as the same shall become payable according to the tenor of said bonds, and of the coupons thereto annexed, have made and delivered to John Ferguson and Samuel J. Tilden, of the City and State of New York, Trustees, three several deeds of trust or mortgage, known, respectively, as the First, Second and Third Mortgages, bearing date, respectively, on the first day of March, in the year one thousand eight hundred and sixty-two, and conveying to the said Trustees, in the manner and for the purposes, and upon the trusts therein specified, all and singular, the railway and the premises, franchises, property, and things therein and hereinafter specified, mentioned and described.

Now THIS INDENTURE WITNESSETH, That the parties of the first part, in consideration of the premises, and of one dollar to each of them in hand paid, the receipt whereof is hereby acknowledged, have granted, bargained and sold, and do, by these presents, grant, bargain, sell, convey and transfer unto the said party of the second part, all the right, title and interest of them, the parties of the first part, and of any or either of the said parties, acquired by virtue of a deed, bearing date the eighteenth day of February, 1862, and made to the said James F. D. Lanier, Samuel J. Tilden, Louis H. Meyer, J. Edgar Thomson and Samuel Hanna, parties hereto of the first part, by John Ferguson and Thomas E. Walker, both of the City and State of New York, Trustees and Special Master Commissioners, in pursuance of decrees of the Circuit Courts of the United States, for the Northern District of Ohio, the Western District of Pennsylvania, the District of Indiana, and the Northern District of Illinois, in causes in Chancery in the said Courts then depending, wherein Charles Moran and others were complainants, and the Pittsburgh, Fort Wayne and Chicago Railroad Company, and others, were defendants, or acquired by virtue of a deed bearing date the twenty-sixth day of February, 1862, made to the said James F. D. Lanier, Samuel J. Tilden, Louis H. Meyer, J. Edgar Thomson and Samuel Hanna, parties hereto of the first part, by the Pittsburgh, Fort Wayne and Chicago Railway Company, pursuant to the aforesaid decrees, of, in, and to all and singular, the continuous Railway, extending from its eastern terminus in Pittsburgh, in the State of Pennsylvania, to Chicago, in the State of Illinois, and commonly known as the PITTSBURGH, FORT WAYNE AND CHICAGO RAILROAD OR RAILWAY, including all the railways, ways and rights of way, depot grounds and other lands; all tracks, bridges, viaducts, culverts, fences, and other structures; all depots, station-houses, engine-houses, car-houses, freight-houses, wood-houses and other buildings; and all machine shops and other shops, held or acquired for use in connection with the said railway, or the business thereof; and including

also, all locomotives, tenders, cars and other rolling stock
or equipment ; and all machinery, tools, implements, fuel and
materials for the constructing, operating, repairing, or replacing
the said railway, or any part thereof, or any of its equipments or
appurtenances ; and also, all franchises connected with or relating to the said railway, or the construction, maintenance, or use
thereof; and all property, franchises, rights and things, of whatever name or nature, which were conveyed by the aforesaid
deeds of the said Trustees and Master Commissioners, and of the
said Company, to the parties of the first part hereto; *subject,
nevertheless*, to the aforesaid certain deed of trust or mortgage,
bearing date on the first day of March, 1862, made by the said
James F. D. Lanier, Samuel J. Tilden, Louis H. Meyer, J. Edgar
Thomson and Samuel Hanna, parties of the first part, to the
said John Ferguson and Samuel J. Tilden, Trustees, creating a
first lien upon the property therein and herein mentioned, for the
purpose of securing the payment of bonds of the said Company,
known as First Mortgage Bonds, amounting in the aggregate to
five millions and two hundred and fifty thousand dollars; and
subject also, to the aforesaid certain other deed of trust or mortgage, bearing date on the said first day of March, 1862, made by
the said James F. D. Lanier, Samuel J. Tilden, Louis H. Meyer, J.
Edgar Thomson and Samuel Hanna, parties of the first part, to
the said Trustees, creating a second lien upon the aforesaid property, for the purpose of securing the payment of bonds of the
said Company, known as Second Mortgage Bonds, amounting in
the aggregate to five millions and one hundred and sixty thousand dollars ; and *subject also*, to the aforesaid certain deed of
trust or mortgage, bearing date on the said first day of March,
1862, made by the said James F. D. Lanier, Samuel J. Tilden,
Louis H. Meyer, J. Edgar Thomson and Samuel Hanna, parties
of the first part, to the said Trustees, creating a third lien upon
the aforesaid property, for the purpose of securing the payment
of bonds of the said Company, known as Third Mortgage Bonds,
amounting in the aggregate to two millions of dollars; and *subject, also*, as to so much of the said railroad, lying between the Federal Street Station, in Alleghany City, in the State of Pennsylvania, and the passenger depot of the Pennsylvania Railroad, on Liberty street, in Pittsburgh, in said State, and of the other property connected therewith, as is embraced within a certain mortgage, made by the Ohio and Pennsylvania Railroad Company to

Thomas T. Firth, of the City of Philadelphia, and Reuben Miller, Jr., of the City of Pittsburgh, bearing date May 6th, 1856, to the lien created by the said mortgage; and *subject also*, as to the rights of way, depot grounds, lots and premises, in the City and vicinity of Chicago, purchased during the pendency of the said suit, in which Charles Moran and others are complainants, and the said Pittsburgh, Fort Wayne and Chicago Railroad Company, and others, are defendants, as aforesaid, by the agency of the Receiver in the said causes, to the lien of the vendors for the purchase money of the same; and subject *also*, as to any real estate included in the aforsaid sale by the said Trustees and Master Commissioners, to the lien, if any such exists, of any vendor or former owner of said real estate, not otherwise provided, to be paid by the decrees or orders in the said causes: *Provided, nevertheless*, and it is the true intent and meaning of these presents, that nothing herein contained shall be construed to express or imply any covenant, by the parties of the first part, or either of them, but that this instrument shall operate to convey, in behalf of the said parties, all the estates and interest in the railway and appurtenances, property, rights, franchises and things hereinbefore described, which the said parties, or either of them, might hold by virtue of the aforesaid conveyances, and which the said parties, each for himself or herself, and not one for the other, can lawfully convey, and no more; and that the said estates and interests are hereby charged with and shall pass, by virtue of these presents, subject to the payment of all liabilities incurred in respect to the said railway or its business, by the said parties of the first part, during their possession of the said rail, way. *Together* with all and singular the tenements, hereditaments and appurtenances thereunto belonging, or in anywise appertaining, and the reversions, remainders, tolls, incomes, rents, issues, and profits thereof; and also all the estate, right, title, interest, dower and right of dower, property, possession, claim and demand whatsoever, as well in law as in equity, of the said parties of the first part, or of either of them, of, in, and to the same, and any and every part thereof, with the appurtenances. *To have* and *to hold* the above-described premises, subject, as aforesaid, unto the said party of the second part, and its successors and assigns, to the only proper use and behoof the said party of the second part, and its successors and assigns, for ever.

And the said party of the second part, for itself and its successors, in consideration of the premises, and of one dollar to it in hand paid by the said parties of the first part to these presents, the receipt whereof is hereby acknowledged, hereby covenants and agrees, to and with the said James F. D. Lanier, Samuel J. Tilden, Louis H. Meyer, J. Edgar Thomson and Samuel Hanna, parties of the first part to these presents, and the survivors and survivor of them, and the executors, administrators and assigns of such survivor, that it, the said party of the second part, and its successors, shall and will at all times hereafter perform and keep all and every the conditions, covenants, agreements and provisions contained in said several deeds of trust or mortgage, or either of them, hereinbefore mentioned, to be by the said party of the second part performed or kept.

And the said party of the second part, for itself and its successors, in consideration of the premises, and of one dollar to it in hand paid by the said parties of the first part, the receipt whereof is hereby acknowledged, further covenants and agrees, to and with the said James F. D. Lanier, Samuel J. Tilden, Louis H. Meyer, J. Edgar Thomson and Samuel Hanna, parties of the first part, and the survivors and survivor of them, and the executors, administrators and assigns of such survivor, that whenever and as often as the said party of the second part, or its successors, shall hereafter acquire any lands, or any equipment, or any other property or things of whatever name, or nature for use in connection with the railway hereinbefore mentioned, or any part thereof, or of any of its equipments or appurtenances, or shall acquire any franchises, including every franchise to be a corporation, which may be hereafter granted to the said Company, the said party of the second part, and its successors, shall and will acquire, possess and hold the same, and each and every thereof, and will likewise hold the franchises to be a corporation, heretofore granted to the said Company, upon the trusts of the three several deeds of trust or mortgage hereinbefore mentioned, until conveyances thereof, in pursuance of the covenant next hereinafter contained, shall be duly made and delivered to the Trustees of the said several deeds of trust or mortgage, respectively.

And the said party of the second part, for itself and its successors, in consideration of the premises, and of one dollar to it in hand paid by the said parties of the first part, the receipt whereof is hereby acknowledged, hereby further covenants and agrees, to and with the said James F. D. Lanier, Samuel J. Tilden, Louis H. Meyer, J. Edgar Thomson and Samuel Hanna, parties of the first part, and the survivors and survivor of them, and the executors, administrators and assigns of such survivor, that the said party of the second part, and its successors, shall and will, from time to time, and at all times hereafter, and as often as thereunto requested by the Trustees, or by the surviving or continuing Trustee, or their or his successors or successor, of either of the said several deeds of trust or mortgage respectively hereinbefore mentioned, execute, deliver and acknowledge all such further deeds, conveyances and assurances in the law, for the better assuring unto the said Trustees, or to the surviving or continuing Trustee, or their or his successors or successor in the trust created by the said deeds of trust, upon the trusts therein expressed, the railway, equipments and appurtenances hereinbefore mentioned or intended so to be, and all other property and things, whatsoever, which may be hereafter acquired for use in connection with the same, or any part thereof, and all franchises now held or hereafter acquired, including the franchise to be a corporation, as by the said Trustees, or by the surviving or continuing Trustee, or their or his successors or successor, or by their or his counsel learned in the law, shall be reasonably advised, devised or required.

And the said party of the second part, for itself and its successors, in consideration of the premises, and of one dollar to it in hand paid by the said parties of the first part, further covenants and agrees, to and with the said James F. D. Lanier, Samuel J. Tilden, Louis H. Meyer, J. Edgar Thomson, and Samuel Hanna, parties of the first part, and the survivors and survivor of them, and the executors, administrators, and assigns of such survivor, that the said party of the second part, and its successors, shall and will, at all times hereafter, keep open an office or agency in the City of New York, for the payment of interest and principal of the bonds of the Pittsburgh, Fort Wayne and Chicago Railway Company, as the same shall become payable, according to the tenor of the said bonds or of the coupons thereto annexed, as recited and provided in the three several deeds of trust or mortgage hereinbefore mentioned; for the redemption of the principal of the said bonds, as provided in said deeds of trust or mortgage; for the transfer of the capital stock of the said Company; for the registration of the bonds and bondholders, and for such other business of the said Company as in and by the said three several deeds of trust or mortgage, or either of them, is provided to be done in the said City of New York.

> IN WITNESS WHEREOF, the parties of the first part have hereunto set their hands and seals ; and the party of the second part has caused its corporate seal to be affixed to these presents, and the same to be attested by the signatures of its President and Secretary, to evidence the acceptance, by the said party of the second part, of the foregoing conveyance, and its execution of the covenants therein contained.

JAMES F. D. LANIER,   [L. S.]
MARY M. LANIER,   [L. S.]
SAMUEL J. TILDEN,   [L. S.]
LOUIS H. MEYER,   [L. S.]
ANN CHARLOTTE MEYER, [L. S.]
J. EDGAR THOMSON,   [L. S.]
LAVINIA F. THOMSON,   [L. S.]
SAMUEL HANNA,   [L. S.]
ELIZA HANNA,   [L. S.]

Signed, sealed and delivered in }
   the presence of   }

    JAMES P. SINNOTT and
    JOHN L. DRUMMOND.

{ Seal Ry }
{ Co. }

W. H. BARNES, *Secretary.*   G. W. CASS, *President.*

[*Note.*—The foregoing deed was duly acknowledged, by the grantors therein named, at the City of New York, before CHARLES NETTLETON, Commissioner for the several States, and Notary Public, on or about the twenty-third day of April, 1862, and was, subsequently, also duly acknowledged by the Railway Company, and is on record.]

6. DEED OF FURTHER ASSURANCE.

THIS INDENTURE, made this twentieth day of September, in the year of our Lord, one thousand eight hundred and sixty-two, between the PITTSBURGH, FORT WAYNE AND CHICAGO RAILWAY COMPANY of the first part, and JOHN FERGUSON and SAMUEL J. TILDEN, of the City and State of New York, of the second part :

Whereas, in and by a certain indenture, bearing date the second day of March, A. D. 1862, made and delivered by and between James F. D. Lanier, and Mary M. Lanier, his wife, Samuel J. Tilden, bachelor, and Louis H. Meyer and Ann Charlotte Meyer, his wife, of the City and State of New York, J. Edgar Thomson, and Lavina F. Thomson, his wife, of the City of Philadelphia and State of Pennsylvania; and Samuel Hanna, and Eliza Hanna, his wife, of the town of Fort Wayne and State of Indiana, of the first part, and the Pittsburgh, Fort Wayne and Chicago Railway Company of the second part, the said Company covenanted and agreed to execute, deliver and acknowledge all such further deeds, conveyances and assurances in the law, for the better assuring unto the trustees of the three several deeds of trust or mortgage in the said indenture mentioned, upon the trusts in the said deeds of trust or mortgage expressed, the railway, its equipments and appurtenances therein mentioned, or intended so to be, and all other property and things whatsoever which might be thereafter acquired for use in connection with the same, or any part thereof, and all franchises then held or thereafter acquired, including every franchise to be a corporation as by the said trustees should be required.

And whereas, the said parties of the second part to these presents, as Trustees of the said three several deeds of trust or mortgage, have requested the said party of the first part to these presents to execute, deliver and acknowledge to the said parties of the second part, as such Trustees, a deed of trust or mortgage in pursuance of the said covenant.

Now this Indenture witnesseth, That the party of the first part, in consideration of the premises, and of one dollar to it in hand paid, the receipt whereof is hereby acknowledged, has granted, bargained and sold, and does by these presents grant, bargain, sell, convey and transfer, unto the parties of the second

part, all and singular, the continuous Railway, extending from its
eastern terminus in Pittsburgh, in the State of Pennsylvania, to
Chicago, in the State of Illinois, and commonly known as the
PITTSBURGH, FORT WAYNE AND CHICAGO RAILROAD OR RAILWAY; including all the railways, ways and rights of way, depot grounds
and other lands, all tracks, bridges, viaducts, culverts, fences and
other structures, all depots, station-houses, engine-houses, car-
houses, freight-houses, wood-houses and other buildings, and all
machine shops and other shops, held or acquired for use in connection with the said railway or the business thereof; and including, also, all locomotives, tenders, cars and other rolling stock or
equipment, and all machinery, tools, implements, fuel and materials, for the constructing, operating, repairing or replacing the
said railway, or any part thereof, or any of its equipments or appurtenances; and also all franchises connected with or relating
to the said railway, or the construction, maintenance or use
thereof; and also all the property, franchises, rights and things,
of whatever name or nature, which were conveyed by the
aforesaid indenture, to the party of the first part hereto;
and also all lands, equipments, and other property and things, of
whatever name or nature, which have been acquired, or which
may be hereafter acquired, for use in connection with the said
railway, or any part thereof, or any of its equipments or appurtenances; and all franchises now held, or which may be hereafter
granted to or acquired by the said Company, including all franchises to be a corporation; *subject, nevertheless,* to the three several deeds of trust or mortgage in said indenture mentioned, and
to the other liens in the said several deeds of trust or mortgage,
and in the said indenture, or either of them, mentioned: *Together*
with all and singular the tenements, hereditaments, and appurtenances thereunto belonging, or in any wise appertaining; and
the reversions, remainders, tolls, incomes, rents, issues, and profits thereof; and also all the estate, right, title, interest, property,
possession, claim, and demand whatsoever, as well in law as in
equity, of the said party of the first part, of, in, and to the same,
and any and every part thereof, with the appurtenances. *To*
have and to hold the above-described premises and appurtenances,
subject as aforesaid, unto the said parties of the second part, as
joint tenants, and not as tenants in common, and to the survivor
of them, and to the heirs and assigns of such survivor, to the
only proper use and behoof of the said parties of the second

part, and of the survivor of them, and of the heirs and assigns of such survivor: *In trust*, nevertheless, for the purposes expressed, and upon the trusts declared in the said three several deeds of trust or mortgage in the said conveyance mentioned, according to the priorities in and by the said deeds of trust or mortgage, respectively, established.

> IN WITNESS WHEREOF, the party of the first part has caused its corporate seal to be hereunto affixed, and the same to be attested by the signatures of its President and Secretary, and the parties of the second part have also set their respective hands and seals, for the purpose of evidencing their acceptance of the trust hereby created, on the day and year first above written.

G. W. CASS, *President.*

W. H. BARNES, *Secretary.*

[Corp. Seal.]

Sealed and delivered }
in presence of }

PROCEEDINGS

TO

MODIFY THE CONDITIONS OF MORTGAGES

SO AS TO ADMIT OF AN

INCREASE OF CAPITAL.

PROPOSITIONS MADE BY THE BOARD TO THE BONDHOLDERS BY RESO-
LUTIONS PASSED APRIL 6, 1864.

Resolved, That the Pittsburgh, Fort Wayne and Chicago Railway Company, in conformity with a resolution of the stockholders, adopted at their regular annual meeting, held in the
City of Pittsburgh on the 16th March, 1864, hereby proposes to
the First and Second Mortgage Bondholders, when assembled in
meetings pursuant to the respective deeds of trust, on Thursday,
the 7th of April, 1864, the following

FINANCIAL PROGRAMME :

Whereas, the said first and second deeds of trust or mortgage
each contain a provision in these words :

" ARTICLE EIGHTH.—The aggregate amount of the capital
stock of the said Pittsburgh, Fort Wayne and Chicago Railway
Company, outstanding at any one time, shall never exceed in par
value six millions and five hundred thousand dollars, unless the
holders of bonds secured by these presents shall have, by a vote
at a meeting duly held, expressly consented to such increase."

It is proposed that the said bondholders shall, in meetings duly held in conformity to Article Twelfth of the said deeds, consent to such increase of the capital stock of the said Company as shall be necessary for the purpose of construction connected with the said railway, to wit, providing additional equipment, machinery, and implements, and such buildings, grounds and other improvements as are properly appurtenant thereto, and are needful facilities to its business, and such portions of a double track as the traffic shall from time to time require.

That such consent be given upon the following conditions:

First.—That all moneys raised by such increase of the capital stock shall be inviolably appropriated to the aforesaid purpose; that the sales or disposition of said stock from time to time, and the application of the moneys, be made under the general supervision of James F. D. Lanier, J. Edgar Thomson, Springer Harbaugh, the present Finance Committee of the Company, and Samuel J. Tilden, and Louis H. Meyer, Directors, resident in New York, or a majority of them, or by agents appointed by them, and that in case of any misapplication of the said moneys, the said Committee shall have power to suspend any further issues of stock than shall at that time have been actually made, and to revoke this consent as to such further issues.

Secondly.—That the total issues for the year 1864 shall not exceed $3,500,000; and that no issues for any future year shall be made, except upon the detailed estimates submitted by the Board of Directors of the amount of money necessary for the work of the year, and after the adoption of a resolution authorizing the same at the annual meeting of the stock and bondholders.

Thirdly.—That this consent shall not take effect, except cotemporaneously with the agreement hereinafter specified, between the said Company and the said First and Second Mortgage Bondholders, and as part of one entire financial system.

Whereas, the said first and second deeds of trust each contain substantially the following provision:

"ARTICLE SIXTH.—In any six months, the first such period commencing on the first day of January, 1862, in which the net earnings of the said railway shall exceed the amount necessary to pay the interest upon all the bonds of the said

Company, secured by these presents, or by either of the two several trust deeds, bearing even dates herewith. and then outstanding, including the contribution to the special sinking fund for the six per cent. bonds herein provided, and a dividend of three per cent. upon six millions and five hundred thousand dollars of capital stock, such surplus shall be reserved, and shall, within sixty days after the expiration of the said six months, be paid over to the trustees, as a sinking fund for the redemption of the bonds secured by these presents. *Provided*, that after the redemption of bonds amounting in the aggregate to two millions and five hundred thousand dollars, the said sinking fund may be limited to the application of not less than one per cent., per annum, upon the aggregate of bonds outstanding at the time of such limitations ; and the said sinking fund may at any time hereafter be varied in its amount, by agreement between the said Company and the holders of the bonds secured hereby, acting by a majority in interest ; and, with the assent of the trustees, the said surplus, or any part thereof, may, as the same shall from time to time arise, be applied to the improvement of the said railway, its equipments and appurtenances.''

It is proposed, that the sinking fund provided in the said several deeds be varied in amount, by agreement between the said Company and the holders of the respective bonds, acting by a majority in interest ; and be reduced and limited to the amount of one per cent. in each year, beginning on the first day of January, 1864, upon the whole original issue of the said bonds ; or, $52,500, for the First Mortgage Bonds, and $51,600 for the Second Mortgage Bonds, in addition to the interest upon the bonds which shall have been purchased in for the sinking fund ; so that the sum to be applied in each year to the payment of interest and sinking fund, shall be eight per cent. upon the amount of the original issue.

That such agreement be entered into by the bondholders upon the following conditions :

First.—That the payments into the said sinking funds be made in priority to any dividends on the stock.

Secondly.—That the sinking fund for the First Mortgage Bonds be applied to the purchase, by the trustees, of the said bonds, at their market value ; and the sinking fund for the Second Mortgage Bonds be applied to the purchase of the said bonds, by the trustees, at their market value; such application to be made in such manner and subject to such qualifications as may be agreed upon between the trustees and the Company.

Thirdly.—That the said Company shall, by an instrument, in a form to be approved by the trustees under the said first and second trust deeds, and duly executed and delivered to the said trustees, waive, relinquish, and extinguish any and all right, which the said Company has or may have after July 1st, 1867, to redeem any of the said bonds, or to require the holders then to accept in exchange therefor six per cent. bonds, by virtue of a clause contained in the said bonds, or of Article Seventh of the said several deeds of trust; and shall cause to be stamped, without expense to the holder, on any of the bonds which may be presented for the purpose, an endorsement in evidence of such agreement, in a form to be approved as aforesaid; and shall also in like manner confer upon all holders of the First and Second Mortgage Bonds the right to vote at all corporate meetings at the rate of one vote for each one hundred dollars of the par value thereof, so far as may be done under existing or future legislation.

Fourthly.—That the said Company shall, by an instrument, in a form approved as aforesaid, and executed and delivered as aforesaid, agree to pay the interest, accruing subsequently to January 1st, 1864, upon the Third Mortgage Bonds, semi-annually, on the first days of April and October in each year.

Resolved, That in the event that the bondholders shall agree to the foregoing propositions, they be requested, to authorize their trustees to execute and to accept all instruments necessary or proper to evidence the said consent and agreement; and that the acceptance by the said trustees of instruments, executed by the said Company, for the purposes aforesaid, be deemed the acceptance of the bondholders, respectively, and evidence of the compliance of the Company with all conditions necessary to give effect to the consent and agreement proposed to be given and made by the bondholders.

Resolved, That the President and Vice-President of this Company be authorized, to cause to be prepared, executed and delivered, in behalf of this Company, all instruments which they may deem necessary or proper, to carry into effect the foregoing proposition; and to cause the said instruments to be sealed by the common seal of this Company, and the same to be attested by the President and Secretary.

Resolved, That the President be requested to appear, in person, at the bondholders' meeting, on the 7th inst., to lay before them the foregoing order and resolutions, and to ask their concurrence therein.

MEETING

OF THE

First Mortgage Bondholders

OF THE

PITTSBURGH, FORT WAYNE AND CHICAGO RAILWAY CO.,

Held at the office of Winslow, Lanier & Co., No. 52 Wall street, in the City of New York, on Thursday, the 7th day of April, 1864, at 11 o'clock, A. M.

The meeting was called to order by S. J. Tilden, Trustee; and organized by the appointment of Wm. B. Ogden, Chairman; Louis H. Meyer and Wm. H. Barnes, Secretaries.

The call of the meeting, as published in the newspapers, was then read, viz.:

Notice to the First and Second Mortgage Bondholders of the Pittsburgh, Fort Wayne and Chicago Railway Company.

The Board of Directors of the Pittsburgh, Fort Wayne and Chicago Railway Company have requested the undersigned, to call meetings of the First Mortgage and Second Mortgage Bondholders, in pursuance of the provisions of the deeds of trust, for the purpose of considering certain modifications of the present financial system of the Company, to which the assent of a majority, in interest, of each of those classes of bondholders, by a vote at a meeting, is necessary. The principle of the proposed measures was submitted to the Annual Meeting of the Corporation, held at Pittsburgh, on the 16th inst., at which were represented a majority of the stockholders, and of each class of the bondholders, and was unanimously approved.

The undersigned, therefore, in exercise of the authority conferred upon them by Article 12 of the First and Second Deeds of Trust, give notice, that meetings of the First and Second Mortgage Bondholders will be held, at the banking house of

Winslow, Lanier & Co., No. 52 Wall Street, in the City of New York, on Thursday, the 7th day of April, 1864, at 11 A. M., for the purpose of considering the aforesaid measures.

At such meeting, the bondholders, if registered, may vote, in person or by proxy.

Registration on the voting register can be made, by any holder, at the office or agency of the Company, at the banking house of Winslow, Lanier & Co., where the books are now open for that purpose, on presentation by the holder of his bonds, or on his filing the written order of the person last registered as the holder.

The registration was originally in the names of the persons to whom the bonds were issued. In cases in which the bonds have since changed hands, the present holder can vote only by registering himself, or upon the proxy of the person last registered.

A general attendance of the bondholders is earnestly requested, as well to act on the questions aforesaid, as to perfect the organization contemplated in Article 12 of the Deeds of Trust.

The meetings of the First and Second Mortgage Bondholders will be held separately, but at the same time and place, and the holders of the Third Mortgage Bonds are invited to be present, though their action is not required by the trust deeds.

The resolutions adopted at the annual meeting, and the resolution of the Board of Directors, requesting a call of the meetings of the bondholders, are subjoined.

JOHN FERGUSON,
SAMUEL J. TILDEN,
Trustees for the First and Second Mortgage Bondholders.

RESOLUTIONS

Recommended by the Board of Directors to the stockholders, for adoption at the annual meeting, March 16, 1864:

Resolved, That it is expedient, that expenditures for construction, equipment and objects pertinent thereto, involving new capital, should, in the main, be provided by an increase of capital stock of the Company.

Resolved, That the increase of capital stock should be made solely for the purpose of building and completing a double track,

or such part thereof as may be expedient, and for equipping the
road and providing such additional rolling stock, machinery, appurtenances and other facilities as may be necessary to properly
do all the business which may offer, and for no other purpose, and
should only be issued from year to year, in such amounts as the
stockholders, at their annual meetings, may decide upon. Their
decision shall be based upon detailed estimates, made by the
Board of Directors, of the amount of money necessary for the
work of the year; and that the new issue of stock shall only be
sold after public notice.

Resolved, That the increase of capital for 1864 shall not exceed $3,500,000.

Resolved, That the stockholders hereby recommend, and, as
far as may be necessary, authorize the Board of Directors to make
such agreement or arrangement with the holders of the bonds,
under the trust deeds, as will enable them to increase the capital
stock, as indicated in the preceding resolutions.

Adopted unanimously by vote of $4,562,000, out of $6,500,000
of stock; $3,729,500, out of $5,250,000 of First Mortgage Bonds;
$3,766,000, out of $5,160,000 of Second Mortgage Bonds; and
$1,425,000, out of $2,000,000 of Third Mortgage Bonds, $400,000
not voting because not registered.

Resolutions adopted by the Board of Directors this day:

Resolved, That the Trustees under the mortgages, Samuel J.
Tilden and John Ferguson, be and they are hereby requested to
call, at the earliest moment possible, a meeting of the bondholders, in the City of New York, to act upon the resolutions
adopted by the stockholders, at their annual meeting in the City
of Pittsburgh, on the 16th day of March, 1864, and to obtain, if
possible, the assent of the bondholders thereto.

Offered by L. Meyer, Esq.
Attest:

W. H. BARNES,
Secretary.

Mr. Tilden then offered the following resolutions:

Resolved, That the call of the present meeting, the period
and mode of its notice, and the provisions contained in the

notice of regulations or by-laws, for the present meeting, are hereby approved.

Resolved, That the voting register of the first mortgage bondholders, kept at the office or agency of the Company, in the City of New York, and now submitted to this meeting, is the proper evidence of the right to vote in the present meeting, and the persons in whose names the bonds stand registered are entitled to vote, in person or by proxy, upon the said bonds.

Resolved, That the following rules or by-laws be adopted and established for the government of meetings of the bondholders of the Pittsburgh, Fort Wayne and Chicago Railway Company, until otherwise ordered :

1. Notice of meetings of bondholders may be given by publication, three times a week, for at least two weeks, in three daily newspapers in the City of New York.

2. In the absence from the country, or disability, of either of the Trustees under the mortgages, meetings may be called by the other Trustee. On any requisition of a majority in interest of the bondholders, as shown by the voting register, a meeting shall be called by the Trustees, and on the refusal or neglect of either, by the other, and on the refusal or neglect of both, the meeting may be called by the said majority.

3. A majority in interest of the bondholders shall constitute a quorum.

4. The meetings, when called by the Trustees, or one of them, shall be called to order by either of the Trustees who may be present.

5. Record shall be made of all proceedings of meetings of bondholders, in a book, and certified to by the Chairman and Secretary, and deposited with the Trustees.

6. Any proxy given for a meeting shall be valid at any adjournment thereof, unless the bondholder giving the proxy shall appear in person at such adjourned meeting.

7. The voting register may be closed for not exceeding one week previous to a meeting.

8. The Trustees may make, temporarily, such other regulations, not inconsistent with the foregoing, as shall seem to them necessary.

By consent, the vote on the resolutions and rules and by-laws was deferred.

Mr. CASS, President of the Company, then submitted the resolutions of the Board of Directors, and the "Financial Programme," for the consideration of this meeting, previous to reading which, he made a statement, in brief, of the object of the same. The resolutions and programme read as follows :

Resolved, That the Pittsburgh, Fort Wayne and Chicago Railway Company, in conformity with a resolution of the stockholders, adopted at their regular annual meeting, held in the City of Pittsburgh, on the 16th March, 1864, hereby proposes to the first and second mortgage bondholders, when assembled in meetings pursuant to the respective deeds of trust, on Thursday, the 7th of April, 1864, the following

FINANCIAL PROGRAMME:

Whereas, the said first and second deeds of trust or mortgage each contain a provision in these words :

" ARTICLE EIGHTH.—The aggregate amount of the capital stock of the said Pittsburgh, Fort Wayne and Chicago Railway Company, outstanding at any one time, shall never exceed, in par value, six millions and five hundred thousand dollars, unless the holders of bonds, secured by these presents, shall have, by a vote at a meeting duly held, expressly consented to such increase."

IT IS PROPOSED, that the said bondholders shall, in meetings duly held in conformity to Article Twelfth of the said deeds, consent to such increase of the capital stock of the said Company as shall be necessary for the purpose of construction connected with the said railway, to wit, providing additional equipment, machinery and implements, and such buildings, grounds and other improvements as are properly appurtenant thereto, and are needful facilities to its business, and such portions of a double track as the traffic shall from time to time require.

That such consent be given upon the following conditions :

First.—That all moneys raised by such increase of the capital stock shall be inviolably appropriated to the aforesaid purpose ; that the sales or disposition of said stock from time to time, and the application of the moneys, be made under the general supervision of James F. D. Lanier, J. Edgar Thomson, Springer

Harbaugh, the present Finance Committee of the Company, and Samuel J. Tilden and Louis H. Meyer, Directors resident in New York, or a majority of them, or by agents appointed by them; and that in case of any misapplication of the said moneys, the said committee shall have power to suspend any further issues of stock than shall at that time have been actually made, and to revoke this consent as to such further issues.

Secondly.—That the total issues for the year 1864 shall not exceed $3,500,000; and that no issues for any future year shall be made, except upon the detailed estimates, submitted by the Board of Directors, of the amount of money necessary for the work of the year, and after the adoption of a resolution, authorizing the same, at the annual meeting of the stock and bondholders.

Thirdly.—That this consent shall not take effect except cotemporaneously with the agreement hereinafter specified, between the said Company and the said first and second mortgage bondholders, and as part of one entire financial system.

Whereas, the said first and second deeds of trust each contain substantially the following provision:

" ARTICLE SIXTH.—In any six months, the first such period commencing on the first day of January, 1862, in which the net earnings of the said railway shall exceed the amount necessary to pay the interest upon all the bonds of the said Company, secured by these presents, or by either of the two several trust deeds, bearing even dates herewith, and then outstanding, including the contribution to the special sinking fund for the six per cent. bonds herein provided, and a dividend of three per cent. upon six millions and five hundred thousand dollars of capital stock, such surplus shall be reserved, and shall, within sixty days after the expiration of the said six months, be paid over to the Trustees as a sinking fund for the redemption of the bonds secured by these presents; *Provided*, that after the redemption of bonds amounting in the aggregate to two millions and five hundred thousand dollars, the said sinking fund may be limited to the application of not less than one per cent., per annum, upon the aggregate of bonds outstanding at the time of such limitation; *and the said sinking fund may, at any time hereafter, be varied in its amount, by agreement between the said Company and the holders of the bonds secured hereby, acting by a majority in interest;* and, with the assent of the Trustees, the said surplus, or any part thereof, may, as the same shall from time to time arise, be applied to the improvement of the said railway, its equipments and appurtenances."

IT IS PROPOSED that the sinking fund provided in the said several deeds *be varied in amount,* by agreement between the said Company and the holders of the respective bonds, acting by a majority in interest; and be reduced and limited to the amount of one per cent. in each year, beginning on the first day of January, 1864, upon the whole original issue of the said bonds; or $52,500 for the First Mortgage Bonds, and $51,600 for the Second Mortgage Bonds, in addition to the interest upon the bonds which shall have been purchased in for the sinking fund; so that the sum to be applied in each year to the payment of interest and sinking fund shall be eight per cent. upon the amount of the original issue.

That such agreement be entered into by the bondholders upon the following conditions:

FIRST.—That the payments into the said sinking funds be made in priority to any dividends on stock.

SECONDLY.—That the sinking fund for the First Mortgage Bonds be applied to the purchase, by the Trustees, of the said bonds, at their market value; and the sinking fund for the Second Mortgage Bonds be applied to the purchase of the said bonds, by the Trustees, at their market value; such application to be made in such manner, and subject to such qualifications, as may be agreed upon between the Trustees and the Company.

THIRDLY.—That the said Company shall, by an instrument, in a form to be approved by the Trustees under the said first and second trust deeds, and duly executed and delivered to the said Trustees, waive, relinquish and extinguish any and all right which the said Company has, or may have after July 1st, 1867, to redeem any of the said bonds, or to require the holders then to accept in exchange therefor six per cent. bonds, by virtue of a clause contained in the said bonds, or of Article Seventh of the said several deeds of trust, and shall cause to be stamped, without expense to the holder, on any of the bonds, which may be presented for the purpose, an endorsement in evidence of such agreement, in a form to be approved as aforesaid; and shall also in like manner confer upon all holders of the First and Second Mortgage Bonds the right to vote, at all corporate meetings, at the rate of one vote for each one hundred dollars of the par value thereof, so far as may be done under existing or future legislation.

FOURTHLY.—That the said Company shall, by an instrument in a form approved as aforesaid, and executed and delivered as aforesaid, agree to pay the interest accruing subsequently to January 1st, 1864, upon the third mortgage bonds, semi-annually, on the first days of April and October, in each year.

Resolved, That, in the event that the bondholders shall agree to the foregoing propositions, they be requested to authorize their Trustees to execute and to accept all instruments necessary or proper to evidence the said consent and agreement; and that the acceptance by the said Trustees of instruments executed by the said Company, for the purposes aforesaid, be deemed the acceptance of the bondholders respectively, and evidence of the compliance of the Company with all conditions necessary to give effect to the consent and agreement proposed to be given and made by the bondholders.

Resolved, That the President and Vice-President of this Company be authorized to cause to be prepared, executed and delivered in behalf of this Company, all instruments which they may deem necessary or proper to carry into effect the foregoing proposition; and to cause the said instruments to be sealed by the common seal of this Company, and the same to be attested by the President and Secretary.

Resolved, That the President be requested to appear, in person, at the bondholders' meeting, on the 7th inst., to lay before them the foregoing order and resolutions, and to ask their concurrence therein.

After full discussion and general approval of the programme by the bondholders present, Mr. Tilden offered the following resolutions:

" *Resolved*, That the First Mortgage Bondholders of the Pittsburgh, Fort Wayne and Chicago Railway Company, duly assembled in meeting, do hereby approve, agree to and adopt the foregoing proposition made to them by the said Company.

Resolved, That the First Mortgage Bondholders of the Pittsburgh, Fort Wayne and Chicago Railway Company, duly assembled in meeting, and acting by a majority in interest, represented in person or by proxy, do hereby consent to the increase of capital stock provided for in the foregoing " Financial Programme," presented by the Company, upon the conditions therein expressed.

Resolved, That the First Mortgage Bondholders, of the Pittsburgh, Fort Wayne and Chicago Railway Company, duly assembled in meeting, and acting by the affirmative vote of a majority in interest of the said bonds, represented in person or by proxy, do hereby consent and agree to the modification of the Sinking Fund specified in the foregoing Financial Programme, upon the conditions therein expressed.

Resolved, that the First Mortgage Bondholders, duly assembled in meeting, and acting by the affirmative vote of a majority in interest, do hereby authorize their Trustees to execute and accept all instruments necessary and proper to evidence the said consent and agreement, and to do all acts and things which shall be, in their judgment, necessary or proper to carry into effect the said programme, and that the acceptance by the said Trustees of instruments, executed by the said Company, for the purposes aforesaid, be deemed the acceptance of the bondholders respectively, and evidence of the compliance of the Company with all conditions necessary to give effect to the consent and agreement proposed to be given and made by the bondholders."

Whereupon it was unanimously ordered, that the vote should be taken upon all the foregoing resolutions and rules, or by-laws and the "Financial Programme" above set forth, and that the vote be taken by each bondholder, or person representing a bondholder, signing his name to lists headed " Aye " and " No ;" the bonds held or represented by him, and the number of bond votes held by each, as ascertained from the voting register by tellers.

On motion, Messrs. L. H. Meyer and W. H. Barnes were appointed to be the tellers to receive, count and make record and return of the votes cast.

The bondholders having voted, the meeting, on motion being made, adjourned until to-morrow morning, at 10 o'clock.

Friday Morning, April 8th, 1864.

Adjourned meeting of the First Mortgage Bondholders of the Pittsburgh, Fort Wayne and Chicago Railway Company.

Pursuant to adjournment, the bondholders met at 10 A. M., Mr. Ogden in the Chair ; L. H. Meyer and W. H. Barnes, Secretaries.

The Secretary read the minutes of the meeting of yesterday, which were approved.

The tellers then presented their returns of the vote had at the meeting yesterday, as follows, viz.:

We, the undersigned, tellers appointed to take the vote of the First Mortgage Bondholders of the Pittsburgh, Fort Wayne and Chicago Railway Company, at a meeting held by them on the 7th inst., do now report, that the First Mortgage Bondholders voted as follows : Ayes, $4,175,000. ; Noes, *none*; and that the First Mortgage Bondholders of the Pittsburgh, Fort Wayne and Chicago Railway Company, by an affirmative vote of $4,175,000.00, out of $5,250,000, did signify their approval of the resolutions, &c., &c., voted upon.

(Signed)   LOUIS H. MEYER,
                             WM. H. BARNES,
                                             Tellers.

Whereupon the Chairman then declared that each of the said resolutions having received an affirmative vote of $4,175,000 out of $5,250,000 of First Mortgage Bonds, being an affirmative vote of more than a majority in interest, the said resolutions, and each of them, and the financial programme contained therein, submitted by the Company, were adopted by the First Mortgage Bondholders.

On motion, the meeting adjourned on call.

At the same time and place the bondholders under the second mortgage met, and had the same proceedings as those of the first mortgage, above recited, excepting the report of tellers and declaration of Chairman, which reads as follows, viz.:

We, the undersigned, tellers appointed to take the vote of the Second Mortgage Bondholders of the Pittsburgh, Fort Wayne and Chicago Railway Company, at a meeting held by them on the 7th inst., do now report that the Second Mortgage Bond-holders voted as follows: Ayes, $3,803,000; Noes, *none;* and that the Second Mortgage Bondholders of the Pittsburgh, Fort Wayne and Chicago Railway Company, by an affirmative vote of $3,803,000, out of $5,160,000, did signify their approval of the resolutions, &c., &c., voted upon.

(Signed,)   LOUIS H. MEYER,
                    WM. H. BARNES,
                         Tellers.

Whereupon the Chairman then declared that each of the said resolutions having received an affirmative vote of $3,803,000, out of $5,160,000, of Second Mortgage Bonds, being an affirmative vote of more than a majority in interest, the said resolutions, and each of them, and the " Financial Programme " contained therein, submitted by the Company, were adopted by the Second Mortgage Bondholders.

On motion, the meeting adjourned on call.

(Signed,)   LOUIS H. MEYER,
                    WM. H. BARNES,
                        Secretaries.

CONTRACT WITH MORTGAGE TRUSTEES.
MODIFYING SINKING FUNDS, AND ALLOWING
INCREASE OF CAPITAL STOCK.

This Indenture, made this eighth day of April, in the year of our Lord one thousand eight hundred and sixty-four, between the PITTSBURGH, FORT WAYNE AND CHICAGO RAILWAY COMPANY, of the first part, and JOHN FERGUSON and SAMUEL J. TILDEN, of the City and State of New York, of the second part, witnesseth:

Whereas, the said Pittsburgh, Fort Wayne and Chicago Railway Company, party of the first part, is a corporation duly formed and organized under laws of the States of Pennsylvania, Illinois and Indiana, respectively;

And whereas, heretofore, to wit, on or about the first day of March, 1862, the said Company issued certain bonds, known respectively, as its First, Second, and Third Mortgage Bonds;

And whereas, the said bonds were so issued upon the purchase of the railway now owned by said Company, from James F. D. Lanier, Samuel J. Tilden, Louis H. Meyer, J. Edgar Thomson and Samuel Hanna, in evidence of a portion of the consideration therefor, and were secured by three several deeds of trust or mortgage, duly executed and delivered by the said James F. D. Lanier, Samuel J. Tilden, Louis H. Meyer, J. Edgar Thomson, and Samuel Hanna, to John Ferguson and Samuel J. Tilden, Trustees, who are the parties of the second part hereto, conveying the Pittsburgh, Fort Wayne and Chicago Railway, and the equipments, appurtenances and things in said deeds described, in the order of priority by the said deeds established;

And whereas, afterwards, to wit, on or about the second day of March, 1864, the said James F. D. Lanier and Mary M. Lanier, his wife, Samuel J. Tilden, bachelor, Louis H. Meyer and Ann Charlotte Meyer, his wife, J. Edgar Thomson and Lavina F. Thomson, his wife, and Samuel Hanna and Eliza Hanna, his wife,

granted and conveyed, to the said Pittsburgh, Fort Wayne and
Chicago Railway Company, the railway, equipments, appurtenances and things in the said several deeds of trust described,
subject to the said several deeds; and the said Company accepted
the same, subject to said deeds, and the liens and priorities thereby
created, and covenanted and agreed, to and with the said James
F. D. Lanier, Samuel J. Tilden, Louis H. Meyer, J. Edgar Thomson and Samuel Hanna, and the survivors and survivor of them,
and the executors, administrators and assigns of such survivor,
that it, the said Pittsburgh, Fort Wayne and Chicago Railway
Company, should and would, at all times thereafter, perform and
keep all and every the conditions, covenants, agreements and
provisions contained in the said several deeds of trust or mortgage, or either of them, to be by the said Company performed or
kept.

And whereas, afterwards, to wit, on or about the twentieth
day of September, 1862, the said Pittsburgh, Fort Wayne and
Chicago Railway Company, in further assurance of the title of
the trustees under the said several deeds of trust, granted and
conveyed to the said parties of the second part hereto, upon the
trusts in the said several deeds of trust expressed, the said Pittsburgh, Fort Wayne and Chicago Railway, its equipments and
appurtenances, and the property and things in the said several
deeds of trust described;

And whereas, the said first and second deeds of trust or
mortgage each contain a provision in these words:

" ARTICLE EIGHTH.—The aggregate amount of capital stock of
" the said Pittsburgh, Fort Wayne and Chicago Railway Com-
" pany, outstanding at any one time, shall never exceed, in par
" value, six millions and five hundred thousand dollars, unless the
" holders of bonds secured by these presents shall have, by a
" vote at a meeting duly held, expressly consented to such in-
" crease."

And whereas, each of the said first and second mortgage
bonds, secured by the first and second deeds of trust, respectively,
contains the following clause, viz.:

" But it is hereby provided, that at any time after the first
" day of July, one thousand eight hundred and sixty-seven, the
" said Company, on any day on which a half-yearly instalment
" of interest shall fall due, may, at their option, redeem at par

"the principal of this bond. And it is further agreed that this
" bond is convertible, at the option of the holder, upon any day
" upon which any such instalment of interest shall become pay-
" able, into a bond, to be issued and secured in the same
" manner as this bond, but bearing interest at the rate of
" six per cent., and irredeemable except by a sinking fund of
" one per cent., per annum, on the whole amount of the said six
" per cent. bonds, which shall have been issued, to be reserved
" and applied in the manner specified in the said deed of trust."

And whereas, the said first and second deeds of trust each
contain, substantially, the following provision :

"ARTICLE SIXTH.—In any six months, the first such period
" commencing on the first day of January, 1862, in which the
" net earnings of the said railway shall exceed the amount neces-
" sary to pay the interest upon all the bonds of the said Com-
" pany, secured by these presents, or by either of the two several
" trust deeds, bearing even dates herewith, and then outstanding,
" including the contribution to the special sinking fund for the
" six per cent. bonds herein provided, and a dividend of three
" per cent. upon six millions and five hundred thousand dollars
" of capital stock, such surplus shall be reserved, and shall,
" within sixty days after the expiration of the said six months, be
" paid over to the Trustees, as a sinking fund for the redemption
" of the bonds secured by these presents; *Provided,* that after
" the redemption of bonds amounting in the aggregate to two
" millions and five hundred thousand dollars, the said sinking
" fund may be limited to the application of not less than one per
" cent., per annum, upon the aggregate of bonds outstanding at
" the time of such limitation ; and the said sinking fund may, at
" any time hereafter, be varied in its amount, by agreement be-
" tween the said Company and the holders of the bonds secured
" hereby, acting by a majority in interest ; and with the assent of
" the Trustees, the said surplus, or any part thereof may, as the
" same shall from time to time arise, be applied to the improve-
" ment of the said railway, its equipments and appurtenances."

And whereas, each of the said bonds also contains the following provision, viz. :

" The person appearing on the Voting Bond Register of the
" said Company as the holder of this bond, at the time of any
" meeting of the stockholders of the said Company, will be en-

" titled to one vote, at such meeting, for every two hundred dol-
" lars of the par amount thereof."

And whereas, at the annual meeting of the stockholders of the said Pittsburgh, Fort Wayne and Chicago Railway Company, held at the City of Pittsburgh, on the sixteenth day of March, 1864, the following resolutions were adopted, viz. :

" *Resolved*, That it is expedient that expenditures for con-
" struction, equipment, and objects pertinent thereto, involving
" new capital, should in the main be provided for by an increase
" of capital stock of the Company.

" *Resolved*, That the increase of capital stock should be made
" solely for the purpose of building and completing a double
" track, or such part thereof, as may be expedient, and for equip-
" ping the road, and providing such additional rolling stock,
" machinery, appurtenances and other facilities as may be neces-
" sary to properly do all the business which may offer, and for no
" other purpose, and should only be issued from year to year, in
" such amounts as the stockholders at their annual meetings may
" decide upon. Their decision shall be based upon detailed esti-
" mates, made by the Board of Directors, of the amount of
" money necessary for the work of the year, and that the new
" issue of stock shall only be sold after public notice.

" *Resolved*, That the increase of capital for 1864 shall not ex-
" ceed $3,500,000.

" *Resolved*, That the stockholders hereby recommend, and,
" as far as may be necessary, authorize the Board of Directors to
" make such agreement or arrangement with the holders of the
" bonds under the trust deeds as will enable them to increase the
" capital stock, as indicated in the preceding resolutions."

And whereas, for the purpose of carrying out the aforesaid recommendations of the stockholders, the following resolutions were, at a meeting of the Board of Directors, unanimously adopted :

Resolved, That the Pittsburnh, Fort Wayne and Chicago Railway Company, in conformity with a resolution of the stockholders, adopted at their regular annual meeting, held in the City of Pittsburgh, on the 16th March, 1864, hereby proposes to the First and Second Mortgage Bondholders, when assembled in meetings pursuant to the respective deeds of trust, on Thursday, the 7th of April, 1864, the following

FINANCIAL PROGRAMME:

Whereas, the said first and second deeds of trust or mortgage each contain a provision in these words:

"ARTICLE EIGHTH. The aggregate amount of the capital "stocks of the said Pittsburgh, Fort Wayne and Chicago Rail- "way Company outstanding, at any one time, shall never exceed, "in par value, six millions and five hundred thousand dollars, "unless the holders of bonds secured by these presents shall "have, by a vote at a meeting duly held, expressly consented to "such increase:

"*It is proposed*, that the said bondholders shall, in meetings "duly held in conformity to article twelfth of the said deeds, "consent to such increase of the capital stock of the said Com- "pany as shall be necessary for the purpose of construction con- "nected with the said railway, to wit, providing additional equip- "ment, machinery and implements, and such buildings, grounds "and other improvements as are properly appurtenant thereto, "and are needful facilities to its business, and such portions of a "double track as the traffic shall from time to time require."

That such consent be given upon the following conditions:

"*First.*—That all moneys raised by such increase of the cap- "ital stock shall be inviolably appropriated to the aforesaid pur- "pose; that the sales or disposition of said stock, from time to "time, and the application of the moneys be made under the "general supervision of James F. D. Lanier, J. Edgar Thomson, "Springer Harbaugh, the present Financie Committee of the "Company, and Samuel J. Tilden and Louis H. Meyer, Direct- "ors, resident in New York, or a majority of them, or by agents "appointed by them, and that, in case of any misapplication of "the said moneys, the said Committee shall have power to sus- "pend any further issues of stock than shall at that time have "been actually made, and to revoke this consent as to such further "issues.

"*Secondly.*—That the total issues for the year 1864 shall not "exceed $3,500,000, and that no issues for any future year shall "be made, except upon the detailed estimates submitted by the "Board of Directors of the amount of money necessary for the "work of the year, and after the adoption of a resolution author- "izing the same at the annual meeting of the stock and bond- "holders.

164

" *Thirdly.*—That this consent shall not take effect, except
" contemporaneously with the agreement hereinafter specified,
" between the said Company and the said First and Second
" Mortgage Bondholders, and as part of one entire financial sys-
" tem.

" *Whereas*, the said first and second deeds of trust each con-
" tain substantially the following provision:

" *Article Sixth.*—In any six months, the first such period com-
" mencing on the first day of January, 1862, in which the net
" earnings of the said railway shall exceed the amount necessary
" to pay the interest upon all the bonds of the said Company,
" secured by these presents, or by either of the two several trust
" deeds, bearing even dates herewith, and then outstanding, in-
" cluding the contribution to the special sinking fund for the six
" per cent. bonds herein provided, and a dividend of three per
" cent. upon six millions and five hundred thousand dollars of
" capital stock, such surplus shall be reserved, and shall, within
" sixty days after the expiration of the said six months, be paid
" over to the trustees, as a sinking fund for the redemption of the
" bonds secured by these presents; *Provided*, that after the re-
" demption of bonds amounting in the aggregate to two millions
" and five hundred thousand dollars, the said sinking fund may
" be limited to the application of not less than one per cent., per
" annum, upon the aggregate of bonds outstanding at the time of
" such limitation; and the said sinking fund may, at any time
" hereafter, be varied in its amount by agreement between the
" said Company and the holders of the bonds secured hereby,
" acting by a majority in interest; and, with the assent of the
" Trustees, the said surplus, or any part thereof, may, as the same
" shall from time to time arise, be applied to the improvement of
" the said railway, its equipments and appurtenances:

" IT IS PROPOSED, that the sinking fund provided in the said sev-
" eral deeds be varied in amount, by agreement between the said
" Company and the holders of the respective bonds, acting by a ma-
" jority in interest, and be reduced and limited to the amount of one
" per cent. in each year, beginning on the first day of January,
" 1864, upon the whole original issue of the said bonds, or
" $52,500 for the first mortgage bonds, and $51,600 for the
" second mortgage bonds, in addition to the interest upon the
" bonds which shall have been purchased in for the sinking fund,
" so that the sum to be applied in each year to the payment of '

" interest and sinking fund shall be eight per cent. upon the
" amount of the original issue.

" " That such agreement be entered into by the bondholders,
" upon the following conditions:

" *First.*—That the payments into the said sinking funds be
" made in priority to any dividends on the stock.

" *Secondly.*—That the sinking fund for the first mortgage
" bonds be applied to the purchase by the trustees of the said
" bonds, at their market value ; and the sinking fund for the
" second mortgage bonds be applied to the purchase of the said
" bonds by the trustees, at their market value ; such application
" to be made in such manner, and subject to such qualifications,
" as may be agreed upon between the trustees and the Company.

" *Thirdly.*—That the said Company shall, by an instrument,
" in a form to be approved by the trustees under the said first
" and second trust deeds, and duly executed and delivered to the
" said trustees, waive, relinquish, and extinguish any and all
" right which the said Company has, or may have after July 1st,
" 1867, to redeem any of the said bonds, or to require the hold-
" ers then to accept in exchange therefor six per cent. bonds, by
" virtue of a clause contained in the said bonds, or of Article
" Seventh of the said several deeds of trust; and shall cause to
" be stamped, without expense to the holder, on any of the bonds
" which may be presented for the purpose, an endorsement in evi-
" dence of such agreement, in a form to be approved as afore-
" said ; and shall also in like manner confer upon all holders of
" the first and second mortgage bonds the right to vote at all
" corporate meetings, at the rate of one vote for each one hun-
" dred dollars of the par value thereof, so far as may be done
" under existing or future legislation.

" *Fourthly.*—That the said Company shall, by an instrument,
" in a form approved as aforesaid, and executed and delivered as
" aforesaid, agree to pay the interest, accruing subsequently to
" January 1st, 1864, upon the third mortgage bonds, semi-annually,
" on the first days of April and October in each year.

" *Resolved,* That in the event that the bondholders shall
" agree to the foregoing propositions, they be requested to
" authorize their trustees to execute and to accept all instruments
" necessary or proper to evidence the said consent and agree-

" ment ; and that the acceptance by the said trustees of instru-
" ments executed by the said Company, for the purposes afore-
" said, be deemed the acceptance of the bondholders respectively,
" and evidence of the compliance of the Company with all con-
" ditions necessary to give effect to the consent and agreement
" proposed to be given and made by the bondholders.

" *Resolved*, That the President and Vice-President of this
" Company be authorized, to cause to be prepared, executed and
" delivered, in behalf of this Company, all instruments, which
" they may deem necessary or proper to carry into effect the
" foregoing proposition ; and to cause the said instruments to be
" sealed by the common seal of this Company, and the same to be
" attested by the President and Secretary.

" *Resolved*, That the President be requested to appear, in
" person, at the bondholders meeting on the 7th inst., to lay be-
" fore them the foregoing order and resolutions, and to ask their
" concurrence therein."

And whereas, at a meeting of the holders of the said first
mortgage bonds, duly convened by the trustees, in conformity
with the provisions of the said first deed of trust, held at the office
of Messrs. Winslow, Lanier & Co., No. 52 Wall street, in the City
of New York, on the seventh day of April, in the year one thousand eight hundred and sixty-four, the President of the said Company presented for the consideration of said meeting the resolutions and " Financial Programme " aforesaid, and asked the concurrence of the said bondholders therein ; and thereupon, after
full discussion and general approval by the said bondholders, the
following resolutions were, on motion, unanimously adopted,
holders of bonds, to the aggregate amount of $4,175,000, present
at said meeting, in person or by proxy, voting in the affirmative,
and there being no vote cast in the negative, viz:

" *Resolved*, That the first mortgage bondholders of the Pitts-
" burgh, Fort Wayne and Chicago Railway Company, duly assem-
" bled in meeting, do hereby approve, agree to and adopt the
" foregoing proposition made to them by the said Company.

" *Resolved*, That the first mortgage bondholders of the Pitts-
" burgh, Fort Wayne and Chicago Railway Company, duly assem-
" bled in meeting, and acting by a majority in interest, repre-
" sented in person or by proxy, do hereby consent to the increase

" of capital stock provided for in the foregoing 'Financial
" Programme' presented by the Company, upon the conditions
" therein expressed.

" *Resolved*, That the first mortgage bondholders of the Pitts-
" burgh, Fort Wayne and Chicago Railway Company, duly assem-
" bled in meeting, and acting by the affirmative vote of a majority
" in interest of the said bonds, represented in person or by proxy,
" do hereby consent and agree to the modifications of the sinking
" fund specified in the foregoing 'Financial Programme,' upon
" the conditions therein expressed."

"*Resolved*, That the first mortgage bondholders, duly assembled
" in meeting and acting by the affirmative vote of a majority in
" interest, do hereby authorize their trustees to execute and ac-
" cept all instruments, necessary or proper to evidence the said
" consent and agreement, and to do all acts and things which
" shall be, in their judgment, necessary or proper to carry into
" effect the said programme, and that the acceptance by the said
" trustees of instruments executed by the said Company, for the
" purposes aforesaid, be deemed the acceptance of the bondhold-
" ers, respectively, and evidence of the compliance of the Com-
" pany with all conditions necessary to give effect to the consent
" and agreement proposed to be given and made by the bond-
" holders."

And whereas, at a meeting of the holders of the said second
mortgage bonds, duly convened by the trustees, in conformity
with the provisions of the said second deed of trust, held at the
office of Messrs. Winslow, Lanier and Company, No. 52 Wall street,
in the City of New York, on the seventh day of April, in the
year one thousand eight hundred and sixty-four, the President of
the said Company presented, for the consideration of the said
meeting, the resolutions and " Financial Programme " aforesaid,
and asked the concurrence of the said bondholders therein ; and
thereupon, after full discussion and general approval by the said
bondholders, resolutions precisely similar to those passed by the
said first mortgage bondholders, and hereinbefore recited, were,
on motion, unanimously adopted, the holders of said second
mortgage bonds, to the aggregate amount of $3,803,000, present
at said meeting, in person or by proxy, voting in the affirmative,
and there being no vote cast in the negative.

Now, THEREFORE, THIS INDENTURE WITNESSETH, That the said party of the first part, for and in consideration of the premises, and of the sum of one dollar, to it paid by the parties of the second part, and in order to carry into operation and effect the agreements so as aforesaid made by it with the said bondholders, has covenanted and agreed, and, by these presents, does covenant and agree, to and with the said parties of the second part, and the survivor of them, and their and his successors and successor in the trust by the said first and second deeds of trust, respectively, created, and to and with each and every holder of the first and second mortgage bonds aforesaid, as follows :

ARTICLE FIRST.—That the said party of the first part has waived, relinquished and extinguished, and by these presents does waive, relinquish and extinguish, any and all right, claim, option or privilege, which the said Company has, or might have, if the propositions aforesaid had not been made to and accepted by the said bondholders, after the first day of July, 1867, to redeem, prior to the first day of July, 1912, any of the said bonds, or to require the holders thereof, as an alternative to such redemption, to accept, in exchange therefor, six per cent. bonds, by virtue of the clause in the said bonds, hereinbefore recited, or of anything in the respective deeds of trust, securing the said bonds, contained ; and that the said bonds, and each and every of them, shall be and remain irredeemable by the said Company until the said first day of July, 1912, anything in the said bonds, or any of them, or in either of the deeds of trust aforesaid, to the contrary notwithstanding ; and that the said Company, without expense to the holder, will cause to be stamped upon, or otherwise affixed or attached to any of the said bonds, which may be presented for that purpose at any office or agency of the said Company, a memorandum or agreement in evidence of such waiver and relinquishment, in the form following :

PITTSBURGH, FORT WAYNE AND CHICAGO RAILWAY COMPANY.

Agreement in Reference to   *Mortgage Bonds.*

The provision of this bond, by which the same may be redeemed by the Company before the maturity thereof, or a six per cent. bond issued in exchange therefor, has been, for a valuable consideration, abrogated and annulled, by an agreement, bearing date April 8th, 1864, between the Company and the trustees under the deed of trust within mentioned, which agreement is hereby adopted and made operative and obligatory between the said Company and the holder hereof; and the right to vote upon this bond has been, by the same agreement, and is hereby enlarged to one vote for every one hundred dollars of the par value thereof, and the sinking fund modified as in the said agreement provided.

> In witness whereof, the said Company has caused its corporate seal to be hereto affixed, and the same to be attested by the signatures of its President and Secretary, the   day of   , A. D. 186   .

President.

Secretary.

And all provisions of the said bonds, and of each and every of them, and all provisions of the aforesaid first and second deeds of trust, whereby the party of the first part hereto has, or might have, any right, claim, option, or privilege to redeem any of the aforesaid bonds before the said first day of July, 1912, or to require any holder thereof, as an alternative to such redemption, to accept in exchange therefor six per cent. bonds, or might create or issue six per cent. bonds under or pursuant to the said deeds of trust, or either of them, are hereby abrogated and annulled.

ARTICLE SECOND.—That the said party of the first part shall and will, for every six months, commencing on the first day of January, 1864, reserve, and shall and will, within sixty days after the expiration of each such period, pay over to the trustees of the first deed of trust aforesaid, the sum of twenty-six thousand two hundred and fifty dollars, being one-half of one per cent. upon the aggregate amount of the original issue of the bonds secured by the first deed of trust aforesaid, and also such additional sum as shall be equal to the semi-annual instalment of interest upon such bonds of the said original issue as shall have been redeemed or purchased by the trustees in the application of the sinking fund hereby provided, so that the aggregate amount applicable in each such period to the payment of inter-

170

est and sinking fund shall be $210,000, being four per cent upon the aggregate amount of the said original issue.

The moneys so paid to the trustees shall be deposited in some depository in the City of New York, which shall, in the judgment of the trustees, be safe, and shall be, from time to time, applied by the trustees to the purchase of bonds of the issue secured by the said first deed of trust, at the market value thereof.

Such purchases shall be made after ten days' notice in one or more newspapers printed in the City of New York, and at the lowest price or prices at which the bonds may be offered, pursuant to such notice, or at such lower price or prices as the trustees may be able to obtain the same after such notice ; or such purchases may be made, in the discretion of the trustees, at the brokers' board, in the City of New York, or at any public sale in said city: *Provided, nevertheless,* that at any time after two-thirds of the whole issue of the said first mortgage bonds shall have been purchased in for the said sinking fund, the trustees shall not be obliged to invest in the said bonds at a premium which they shall deem unreasonable ; and it shall be competent for the said party of the first part, and the holders of the aforesaid bonds, acting by a majority in interest, to enter into any new agreement, which they may deem necessary, for the regulation of the rates at which, and the mode in which, purchases shall be made.

And the bonds so purchased shall be immediately registered as belonging to the said sinking fund, but shall remain in force and be held by the trustees, or placed by them in some safe depository in the said City of New York.

ARTICLE THIRD.—That the said party of the first part shall and will, for every six months, commencing on the first day of January, 1864, reserve, and shall and will, within sixty days after the expiration of each such period, pay over to the trustees of the second deed of trust aforesaid, the sum of twenty-five thousand and eight hundred dollars, being one-half of one per cent. upon the aggregate amount of the original issue of the bonds secured by the second deed of trust aforesaid, and also such additional sum as shall be equal to the semi-annual instalment of interest upon such bonds of the said original issue as shall have been redeemed or purchased by the trustees in the application of the sinking fund hereby provided ; so that the aggregate amount applicable in each such period to the payment of interest and sinking fund shall be $206,400,—being four per cent. upon the aggregate amount of the said original issue.

The moneys so to be paid to the trustees, as last hereinbefore provided, shall be deposited in some depository, in the City of New York, deemed by the said trustees to be safe, and shall be applied by them to the purchase of bonds, secured by the said second deed of trust, at their market value, upon the notice, in the manner and with the powers, authorities and discretion hereinbefore prescribed and granted, in respect to the bonds secured by the said first deed of trust; and the bonds so purchased shall be registered and remain in force, and be held or deposited as in the case of the bonds secured by the said first deed of trust: *Provided, nevertheless,* that at any time after two-thirds of the whole issue of the said second mortgage bonds shall have been purchased in for the sinking fund for the redemption thereof, the trustees shall not be obliged to invest in the said second mortgage bonds at a premium which they may deem unreasonable; and it shall be competent for the said party of the first part, and the holders of the said second mortgage bonds, acting by a majority in interest, to enter into any new agreement which they may deem necessary for the regulation of the rates at which, and the mode in which purchases shall be made.

ARTICLE FOURTH.—It is hereby covenanted, agreed and declared by the party of the first part, that each and every payment herein agreed to be made by the said party of the first part, as provided in articles second and third of these presents, shall be due and payable according to the terms of the said provisions, respectively, without any condition or qualification whatsoever, by reason of anything in the said bonds or the said deeds of trust, or either of them, contained. And it is further covenanted, agreed and declared, that in case of any default in any payment of an instalment into either sinking fund required to be made by said articles second or third of these presents; and if the net earnings of the said railway, during the six months in respect to which such instalment may have accrued, shall have been sufficient to pay said instalment, in addition to paying interest accruing within the said period upon the three several issues of bonds, respectively secured by the first, second and third deeds of trust, exclusively of such of the said issues as may have been purchased in for the sinking funds of the first and second mortgage bonds ; or if, in case of such default, the party of the first part shall have applied, or reserved for the purpose of applying, any portion of the net earnings accruing within the said period to dividends upon the stock of the said Company, or shall have diverted them to any other purpose whatsoever, then, and in either of such cases, the provisions of Article Second, and of subdivision second of Article Thirteenth of the deed of trust securing the bonds, as to which such default may have occurred, shall apply to such default as fully and effectually as if such default had been expressly mentioned therein.

And it is further covenanted, agreed and declared, that so far as the portion of the income of the said sinking fund, arising from bonds purchased and held for its benefit is concerned, all rights and remedies in respect thereto are preserved and continued in full operation, and may be exercised and enforced by the trustees for the benefit of such sinking funds.

ARTICLE FIFTH.—That the person appearing on the voting register of the said Company, as the holder of any of the said first or second mortgage bonds, at the time of any meeting of the stockholders of said Company, shall be entitled to one vote, at such meeting, for every one hundred dollars of the par amount of the first or second mortgage bonds so appearing to be held by him ; and the said Company does by these presents give, grant and confirm unto such person the right to vote at such meetings, at the rate last aforesaid, as fully and effectually, to all intents and purposes, as if the same had been granted or conferred by the said bonds and the said deeds of trust at the original execution thereof.

ARTICLE SIXTH.—That the said party of the first part shall and will pay the interest, accruing subsequently to the first day of January, 1864, upon the third mortgage or "Income" bonds aforesaid, semi-annually, on the first days of April and October, in each year, at the office or agency of the said Company in the City of New York, instead of annually, on the first day of April in each year, as in the said bonds provided, out of the net earnings, as in the said deed of trust defined ; and the obligation of the said bonds and of the deed of trust by which the same are secured, and all the provisions thereof, shall, from and after the first day of January, 1864, apply to the semi-annual payments of interest in the same manner and with the same effect as if the said bonds had originally provided for such semi-annual payments instead of such annual payments of interest.

And the said party of the first part, for and in consideration of the premises, and of the sum of one dollar, to it duly paid by the parties of the second part, further covenants and agrees to and with the said parties of the second part, and the survivor of them, and their and his successors and successor in the trusts by the said first and second deeds of trust respectively created, and to and with each and all the persons who are or may become holders of any of the said first or second mortgage bonds, that the said party of the first part shall and will, at all times hereafter, and as often as thereunto requested by the trustees or trustee for the time then being, under either of the said first or second deeds of trust, execute, deliver and acknowledge all such further written instruments, deeds, conveyances and assurances in the law for the better assuring to the said parties of the second part, and the survivor of them, and their and his successors, and his successor in the trusts by the said first and second deeds of trust respectively created, upon and subject to all and singular the provisions, covenants and agreements in this indenture contained, and upon the trusts in the said deeds of trust expressed, as the same are altered or modified by this indenture, the railway, equipments and appurtenances mentioned or described in the said deeds of trust, and all other property and things whatsoever, which may be hereafter acquired for use in connection with the same, or any part thereof, and all franchises now held or hereafter acquired, including the franchise to be a corporation, as by the said parties of the second part, or the survivor of them, or their successors or successor in the trusts by the said deeds respectively created, or by their or his counsel, learned in the law, shall be reasonably advised, devised or required.

And this Indenture further witnesseth, that, in consideration of the premises, the said parties of the second part, and the holders of bonds secured by the said first and second deeds of trust, respectively, acting by majorities in interest, at meetings duly held, have consented and agreed, and by these presents do consent and agree, that the capital stock of the party of the first part may be increased beyond six millions and five hundred thousand dollars in par value, and to such amount as may be necessary for the purpose of construction connected with the said railway, to wit, providing additional equipment, machinery and implements, and such buildings, grounds and other improvements as are properly appurtenant thereto, and are needful facilities to its business, and such portions of a double track as the business may from time to time require, anything in the said first and second deeds of trust, or either of them, contained, to the contrary in anywise notwithstanding; *Provided always*, that this consent and agreement is upon and subject to the terms and conditions hereinbefore expressed.

In testimony whereof, the party of the first part has caused its corporate seal to be hereunto affixed, and the same to be attested by the signatures of its President and Secretary, and the parties of the second part have hereunto set their hands and seals, in exercise of the powers conferred upon them by the said First and Second Mortgage Bondholders, for the purpose of evidencing the acts and things mentioned in the resolutions of the said bondholders hereinbefore recited, upon the day and year first above written.

Sealed and delivered in the presence of—

CIRCULAR.

To the Bondholders and Stockholders of the Pittsburgh, Fort Wayne and Chicago Railway Company :

The undersigned, Commissioners to carry into effect the recent financial arrangement between the bondholders and stockholders, deem it proper to communicate to you this statement :

1. The applications of earnings of the railroad to improvements of the property, in the nature of construction, have been, according to the books—

During the receivership, under the order entered January 17, 1860	$1,007,650.06
During the six months from Nov., 1861, to May 1, 1862	236,694.60
From May 1, 1862, when the present Company went into possession, to Dec. 31, 1862	970,147.56
(Of which $536,673 were for Depot and Bridge Bonds.)	
During 1863	1,517,162.25
	$3,731,654.47

Which investment is now bearing good interest, in the shape of enlarged earnings, and virtually increases the intrinsic value of the stock full, $3,500,000.

2. This expenditure has been made without increasing the debt or stock of the Company, as re-organized. Net earnings, over interest, supplied a million last year, and large sums before ; and the residue has been provided for by remission of interest, made in conformity to the agreement for re-organization, and by surpluses in the hands of the Bondholders' Committee, resulting from provident adjustments.

3. Before the late financial arrangement, the Company had provided for $1,912,000, to be expended for improvements during the present year, most of which was to have been furnished by net earnings, and the residue from the other resources mentioned. It was not contemplated in the execution of this work, to increase the capital stock, or to incur any additional bonded debt, or any floating debt. No doubt was entertained, that the net earnings, after paying interest, would have been sufficient for these purposes. Neither the officers of the Company, nor the Committee of the Bondholders, felt any hesitation in acting upon this conviction.

4. The purchase and construction of new equipment has formed an important share of these expenditures. Among them was a provision for 95 new locomotives. The cars have been proportionately increased; contracts for iron have also been advantageously made; the track has been greatly improved; it needs still further expenditure; the equipment received is still inadequate to the business which offers. The savings on this work, as compared with the present range of prices, will be nearly two millions of dollars.

5. It was in this condition of things, that the desire of the Board to proceed more rapidly with the improvements contemplated, and the demand of the stockholders for the application of the net earnings to dividends, led to a conference, between these interests and the bondholders, which has resulted in some modifications of the financial plan, fixed in 1859 by the agreement of re-organization.

These modifications are substantially the following :

To the bondholders is accorded—

1. An extinguishment of the right of the Company, after July 1st, 1867, to require them to accept six per cent. bonds, or to receive payment of the principal: a change which leaves all the bonded debt of the Company irredeemable until July 1st, 1912.

2. The establishment of sinking funds for the first and second mortgages, having priority over dividends, and the application of their incomes to the purchase of the bonds, at their market value.

3. Incidentally, an increase of their security, by a large additional expenditure of new capital upon the property.

4. An enlargement of the voting power of the first and

second mortgage bondholders, from one vote on every $200 to one vote on every $100, of par value.

5. The payment of interest on the third mortgage bonds, semi-annually, instead of annually.

To the stockholders, upon these conditions, is accorded:

1. The power to provide for all new construction by issues of new capital stock.

2. A release of the provisions of the trust deeds, which require all net earnings, over 6 per cent. on the original $6,500,000 of capital, to be applied as sinking funds to purchase in the bonds, or, in some cases, to new improvements.

The practical result of these measures is, to leave the net earnings in each year, after paying interest and sinking funds, at the disposal of the Company, for dividends upon the stock; and the policy may now be deemed to be settled, by the unanimous action of the stockholders and bondholders, to apply such surplus of net earnings to dividends, as far as prudence and sound discretion will warrant.

6. The annual charge for interest and sinking funds will be as follows:

8 per cent. on $5,250,000 of 7 per cent. first mortgage bonds..	$420,000
8 per cent. on $5,160,000 of 7 per cent. second mortgage bonds..................................	412,800
7 per cent. on $2,000,000 of third mortgage or income bonds..................................	140,000
On Bridge bonds, and Chicago depot bonds........	30,950
Total..............................,..................	$1,003,750

7. The net earnings for 1863, as stated in the President's Annual Report, were $2,106,623.18, which would have been sufficient to pay the interest on the bonds and the instalments of the two new sinking funds, and to leave a surplus of $1,103,873.18, which, if the new financial arrangement had then existed, would have been applicable to dividends on $6,500,000 of stock, being over 16.96 per cent.

This was, in the main, without the benefit of the $1,517,162 of new capital expended on the road and equipment, during that year, and with an equipment very inadequate to the business which offered.

The President's estimate of net earnings for the present year, made also in the Annual Report, is $2,500,000, and, in a special report, he stated, that the net earnings for March, after deducting the proportion for interest and the new sinking funds, were over 2½ per cent. on the present amount of the stock.

Thus far, the receipts for April are quite equal to those of March.

8. In judging of the value and productiveness of your property, it is to be noted—

That it is a direct and continuous line, without branches or dependencies, between great industrial and commercial centres ; 468 miles long, every part having a through business and an abundant local traffic, rapidly increasing, with easy grades, and slight curves, more than two-thirds of the distance being straight line, traversing a grain growing region of unsurpassed fertility, as yet but partially developed.

That the investment of capital in bonds and stock is less than that of any other leading line of equal, or even similar, productive capacity, by more than one-third.

That the present affluence of earnings is mainly due to these permanent causes.

If the business capacity of the line be properly sustained and developed, there is no reason to doubt, in the judgment of the undersigned, that it will maintain its dividends, in every condition of circumstances which will allow of dividends in similar enterprises.

It must be borne in mind, also, that as *none of the bonds are payable for* 48 *years, and none of them are convertible into stock,* the whole increase of net earnings, arising from a natural and permanent growth of business, inures to the stock, which forms at present but one-third of the invested capital. The less permanent effect of an inflated currency will operate, during its continuance, in the same manner.

The undersigned, in the exercise of the discretion intrusted to them by the bondholders and stockholders, will not now press the sale of any new issue of stock under the new arrangement; they see no cause for a premature issue. That measure can be deferred until an easier condition of the money market ; and the gradual and general distribution among permanent investors of the present stock, will enable us more nearly to obtain its real value.

Dated New York, April 27, 1864.

> JAMES F. D. LANIER,
> J. EDGAR THOMSON,
> SPRINGER HARBAUGH,
> SAMUEL J. TILDEN,
> LOUIS H. MEYER,
> *Commissioners.*

NOTE.—The following is a copy of the Special Report made by the President, hereinbefore referred to:

NEW YORK, April 8, 1864.

GENTLEMEN : The unanimous action of the stock and bond-holders has liberated the net income of the railway, after paying interest and sinking fund, so as to place it at the disposal of the Board of Directors, for the purposes of dividends, surplus funds, and other objects consistent with the interests of the corporation.

The first and second mortgage bonds, being $\frac{5}{6}$ (five-sixths) of the whole funded debt, were so arranged in monthly instalments of $\frac{1}{12}$ (one-twelfth) each month. Two years having confirmed the theory upon which the plan was adopted, it might now be well, to assimilate to it the plan of paying dividends on the capital stock by paying quarter-annual dividends, instead of annual or semi-annual dividends, as is generally customary with other corporations.

I herewith submit an approximate statement of the earnings, expenses and disbursements for the months of January, February and March past, which is sufficiently near correct to justify the Board acting upon it with a view of making a dividend, if it is thought best to declare one at this time:

Earnings for January $290,675 81
Earnings for February........................ 455,211 02
Earnings for March............................ 602,603 28

Total for three months $1,348,490 11

Deduct operating expenses :

For January	$210,296	21
For February	263,328	22
For March	330,000	00
	803,624	43

Balance...................................$544,865 68

Less interest and sinking fund as follows :

Three months on 1st mortgage	$91,875	00
Three months on 2d mortgage	90,300	00
Three months on 3d mortgage	35,000	00
Three months on Chicago depot bonds.	1,737	50
Three months on interest and sinking fund on Alleghany Bridge bonds...	6,000	00
Sinking fund of one per cent. on 1st and 2d mortgages for three months	26,025	00
	250,937	50

Leaving a balance of......................$293,928 18

Which is applicable to dividends, and subject to the order of the Board. The above balance of unexpended and unappropriated income is equal to a dividend of 4.52 per cent. on the outstanding capital stock.

The small earnings for January were caused, as you are aware, by an almost total suspension of the business of the railway, for about one-fourth of the whole month, arising out of a combination of the locomotive runners to interfere with the police and management of the railway.

It would not be prudent, at this time, to divide all the surplus earnings, as unforeseen contingencies may affect future earnings, and, to enable the Company to make full dividends, in adverse times, a surplus fund ought to be accumulated.

Respectfully submitted.

(Signed)   G. W. CASS,
President.

To the Board of Directors of the Pittsburgh, Fort Wayne and Chicago Railway Company.

Upon the reading of the above report, it was unanimously resolved to declare a dividend of 2½ per cent. for the quarter year, from January 1 to March 31, 1864, payable at the office or agency of the Company, on the 15th day of May next.

AGREEMENT

BETWEEN

THE PITTSBURGH, FORT WAYNE AND CHICAGO RAILWAY COMPANY,

AND

THE CLEVELAND AND PITTSBURGH RAILROAD COMPANY.

THIS AGREEMENT, between the Pittsburgh, Fort Wayne and Chicago Railway Company, of the first part, and the Cleveland and Pittsburg Railroad Company of the second part,

WITNESSETH : *Whereas*, it is deemed, for the common benefit of the parties hereto, that they should enter into the arrangements following :

Now, therefore, the parties of the first and second part, in consideration of the covenants herein contained, to be kept and performed, the one with the other, do hereby stipulate and agree with each other as follows :

FIRST.—This contract shall continue in force for twenty-five (25) years, and may be modified or altered, at any time, by the concurrent action of the Board of Directors of the Companies parties to this agreement.

SECOND.—The aggregate earnings of the two Companies shall be divided between the said Companies, respectively, in the proportion of seventy-three and a half (73½) per centum, to the Pittsburgh, Fort Wayne and Chicago Railway Company, and twenty-six and a half (26½) per centum, to the Cleveland and Pittsburgh Railroad Company. This division shall be made quarterly, as soon as the accounts can be adjusted; but the earnings of the respective Companies, as received, shall be paid into their respective treasuries.

THIRD.—The earnings referred to in the preceding paragraph shall be those arising from transportation of freights, passengers and United States mails, and all other earnings, whatsoever, in operating the roads, inclusive of incomes for the use of engines, rolling stock of one of the parties by the other, or by or on other roads, and of all other property connected with the working of the road, and of rents of track, when the rent is received of and for the Companies' proportion of fares; and also all other incomes derived by rent or otherwise from the use of railroads of others, of depots and tracks, and of all real estate and fixtures used or retained by either parties for depots or other railroad purposes.

FOURTH.—An Executive Committee, composed of the Presidents of the two Companies, and to be enlarged by the addition of one Director from each Board, to be appointed by such Board at any time and as often as such addition may be agreed upon by the concurrent action of the Boards, shall have the power and be charged with the duty of ordering, supervising and directing all necessary contracts or arrangements to be made with connecting roads and lines, parties and individuals, *relative to the transportation of persons and property*, and shall determine the policy to be pursued between them and the parties to this agreement. This Committee shall also be charged with prescribing the manner of keeping the accounts of each company, so far as may be necessary to properly and fairly carry out this contract—and to that end, all officers and agents shall be under the control of the said Committee. This Committee shall meet at stated periods, and as often as may be necessary to promote and protect the interests of the parties hereto, and shall meet on the call of any member of the same. Either Board of Directors may, in the absence or inability of any member of the Committee, representing the said Board, to attend a meeting of said Committee, appoint any other officer or member of such Board, to act in the place of such absent member of the Committee, for the time being. Upon all questions, at any time arising, as to the true, proper, or legal meaning of this contract, or the exercise of powers under the same, or of the policy or details of management for the benefit of the joint interests of the parties, there shall be, on the part of the Committee, a unanimity of opinion before final action, or, in case of disagreement, the Board of Directors of the Pittsburgh, Fort Wayne and Chicago Railway Company shall name a competent, disinterested person, familiar with railway management; and the decision of a majority of the said Committee, including as one of the Committee the person so named in such cases, shall be final as to policy and details of management; and to decide all questions of difference which may have arisen between the members of said Committee; and shall be obligatory upon the parties hereto, and all persons under the control of said Committee. The usual place of the meeting of the Executive Committee shall be in the City of Pittsburgh, but they may, from time to time, appoint other places of meeting. The Committee shall cause a record to be kept, at Pittsburgh, of all their proceedings, including the appointment of all such officers and agents as by this agreement they are authorized to make; also of all contracts or agreements with other Railway Companies, parties, or persons. They shall, from time to time, furnish to each of the respective Boards, copies of all their proceedings, appointments and contracts, and the original shall be open to the inspection of the members of the respective Boards of Directors.

FIFTH.—There shall be appointed by the Executive Committee, a General Superintendent, who shall have full and entire management, under said Committee, of the business of the Roads of the two Companies. The Superintendents of Division of the Pittsburgh, Fort Wayne and Chicago Railway, and the Superintendent of the Cleveland and Pittsburgh Railroad, and all persons on the two Roads subordinate to them, shall be under his appointment and control, as is provided by the organization and by-laws and regulations of the Pittsburgh, Fort Wayne and Chicago Railway Company, as now in force.

The salary and expenses appertaining to the office of General Superintendent shall be paid by the respective Companies, parties hereto, in the proportions fixed in Article Second.

SIXTH.—There shall be one General Freight Agent, and one General Ticket Agent, who shall have charge of the joint and separate business of the said Companies, parties hereto; who shall be appointed and be removable by the Executive Committee, and the payment of whose salaries, office and other expenses appertaining to the respective positions and duties, shall be paid by the parties hereto, in the proportions fixed in Article Second. These officers shall have control of the business, and manner of doing the same, in their respective departments.

All other agents and employees necessary to be employed for the joint interest of the parties hereto, shall be appointed by the Executive Committee, and paid in proportion as aforesaid.

SEVENTH.—Each company shall furnish, and keep in good condition its own machinery, rolling stock, etc., and provide necessary assistants, officers and employees for doing the business of their respective roads; and the machinery and rolling stock of one Company shall not be employed on the road of the other Company, excepting so far as may be necessary to transport the business of the Cleveland and Pittsburgh Railroad Company over the Pittsburgh, Fort Wayne and Chicago Railway, between Pittsburgh and Rochester. Each Company shall, from time to time, increase its machinery, rolling stock, and facilities for business, as the increasing business of the road may require; and in case of their neglect or refusal so to do, it shall be the duty of the Executive Committee to provide, at the expense of the party neglecting or refusing, such machinery, rolling stock and facilities; and they are hereby, irrevocably, constituted agents of such party, for that purpose. The roadway, track and appurtenances of each road to be kept in first class working order.

EIGHTH.—During the existence of this contract, the Cleveland and Pittsburgh Railroad Company shall pay, to the Pittsburgh, Fort Wayne and Chicago Railway Company, for the use of the track, &c., as now used by them, between Rochester and Pittsburgh, seven thousand and eighty-three dollars and thirty-three cents ($7,083.33) per month, monthly, for the use of the said track, &c., between Pittsburgh and Rochester, and also pay one half of the actual expense incurred in keeping that portion of the Pittsburgh, Fort Wayne and Chicago Railway in good repair. In all cases where a more permanent structure is made to replace one worn out, decayed or destroyed, between Pittsburgh and Rochester, the Cleveland and Pittsburgh Railroad Company shall not be required to pay any portion of the excess of cost of such permanent structure over the one formerly in use, but they shall pay one-half of six per centum interest on said excess cost.

NINTH.—All through rates for freight and passengers shall, from time to time, be fixed and maintained by the Executive Committee; they shall, also, from time to time, fix and maintain so much of the local rates of either of the Roads of the parties hereto, as may be necessary in establishing the through rates in connection with any other Roads or mode of communication. They shall establish rates, both local and through, and so cause the business to be done that each Road shall do the business most convenient and natural to it, and which, in joint account, can be most economically done, and produce the largest revenue to the parties hereto.

TENTH.—Each party shall be responsible, for its own damages, losses, injuries, failures, or defaults; and shall protect and indemnify the other party against liability therefor; and shall repay them all outlay and expense made or suffered by them on account thereof.

ELEVENTH.—Each road shall be managed, so far as practicable and consistant with the purpose of this agreement, under its own organization.

TWELFTH.—This contract shall take effect on the first of April, one thousand eight hundred and sixty-three, after approval of the Board of Directors and parties authorized to vote has been had thereto.

Signed, on behalf of the parties hereto, by their respective Presidents, this fifteenth day of December, one thousand eight hundred and sixty-two.

> PITTSBURGH, FORT WAYNE AND CHICAGO
> RAILWAY COMPANY,
>
> By G. W. CASS, *President.*

In presence of—
   W. H. BARNES.

> CLEVELAND AND PITTSBURGH RAILROAD
> COMPANY,
>
> By J. N. McCULLOUGH, *President.*

FRANK LANE.

AMENDMENTS,

MADE TO THE AGREEMENT BETWEEN

THE PITTSBURGH, FORT WAYNE AND CHICAGO RAILWAY COMPANY

AND THE

CLEVELAND AND PITTSBURGH RAILROAD COMPANY.

AMENDMENTS,

made to the Agreement between the Pittsburgh, Fort Wayne and Chicago Railway Company, and the Cleveland and Pittsburgh Railroad Company, bearing date the fifteenth day of December, 1862, and subsequently ratified by the Stockholders of each of said Companies.

Whereas, in view of the prospective and possible changes in the length, connections and business of the Roads of said Companies, and in order to secure the greatest equality which is practicable in the division of the gross earnings of said Companies, the said Pittsburgh, Fort Wayne and Chicago Railway Company, of the first part, and the said Cleveland and Pittsburgh Railroad Company, of the second part, do hereby, in pursuance of the authority conferred upon them, and in consideration of the covenants herein contained, to be kept and performed, one with the other, make the following modifications and amendments to said original agreement :

FIRST.—That whenever the gross earnings of either of said

Companies shall exceed the per cent. fixed and set to such Company by Article Two of said Original Agreement, the Company earning such excess, from and after the first day of April, 1866, shall be allowed to retain fifty per cent. of such excess, after said last mentioned date, as and for the cost of running the same, and the remaining fifty per cent. shall be paid to the other Company.

SECOND.—That the division of gross earnings, as made and fixed by Article Two of said Original Agreement, and the allowance for the excess of gross earnings, as made and fixed by Article One of these Amendments, shall be revised, unless such revision be waived by written notice of both the parties hereto, once in two years. The first revision and alteration shall go into effect on the first day of January, 1867, and each successive revision and alteration shall go into effect at the termination of each successive period of two years thereafter. Said division and allowance, whenever and as often as revised and altered, shall be made and fixed in accordance with the equitable object of these amendments, and with a view to further the same, and said division shall be based upon the relative aggregate earnings of each Road for the two years, terminating with the 31st of December next preceding the date when said revision and alteration shall go into effect, and upon the reasonable and practical prospect of the earnings of the Roads for the next succeeding two years. *Provided, however*, that the division and alteration which shall go into effect on the first day of January, 1867, shall be based upon the relative aggregate earnings of each Road since said Original Agreement went into effect, and upon the reasonable and practical prospect of the earnings of the Roads for the next succeeding two years. If, in order to produce the greatest amount of revenue to the parties hereto, any material portion of the natural or legitimate traffic of either line shall be diverted from said line, and thrown upon the other line, such diversion shall be considered, and due allowance made therefor in any revision of gross earnings which may hereafter be made under these Amendments for the following period of two years.

THIRD.—That the revision and alteration of the division of gross earnings, and of the allowance for the excess of gross earnings, provided for by Article Two of these Amendments, shall be made and fixed by the Executive Committee of the two Companies; but in case the Executive Committee cannot agree, then all questions in difference between them, in respect thereto, shall be submitted to the determination of three arbitrators—one to be selected by the two members of the Executive Committee respectively representing each of the two Companies, and the third to be selected by these two; and the decision of a majority of these arbitrators shall be final and conclusive.

Fourth.—That if the said Executive Committee shall fail to agree upon the division of gross earnings, and upon the allowance for the excess of gross earnings, before the first day of February in each year, when said division and allowance shall be subject to revision and alteration, as hereinbefore provided for, the selection of the two arbitrators to be made by the two members of the Executive Committee, respectively representing the two Companies, shall be made within five days thereafter, and the President of each Company shall forthwith give notice in writing of such selection to the President of the other Company; and the two arbitrators thus selected shall select the third within twenty days after their own selection; and the said arbitrators shall be persons familiar with Railway business. In case either Company shall fail to select an arbitrator, or in case the two arbitrators shall fail to select a third as aforesaid, either Company not in default may, after five days' notice in writing to the other Company, apply to the District Court of Allegheny County, in the State of Pennsylvania, or to any Judge thereof in vacation, to appoint as many arbitrators as shall render the number complete; which Court, or Judge thereof, are hereby authorized and empowered to make said appointment; and said appointment, when made, shall be certified in writing by the Court or Judge making the same; and the decision of these arbitrators shall also be final and conclusive. Until the decision of the arbitrators shall be made and published, no excess of gross earnings shall be paid by either party to the other party, but the decision of the arbitrators, when made, shall relate back to the time when said revision and alterations are by the terms of these Amendments to go into effect.

> In testimony whereof, the parties hereto have caused their names to be subscribed, and their corporate seals to be hereto affixed, by their respective Presi-

dents, this Sixteenth day of February, in the year of our Lord, one thousand eight hundred and sixty-six.

THE PITTSBURGH, FORT WAYNE AND CHICAGO RAILWAY CO.,

By G. W. Cass,   [SEAL.]

President.

Witness present (as to G. W. Cass)
  F. M. Hutchinson.

THE CLEVELAND AND PITTSBURGH RAILROAD CO.,

By J. N. McCullough,   [SEAL.]

President.

Wm. Stewart,
  (As to J. N. McCullough.)

LEASE

OF

THE NEW CASTLE AND BEAVER VALLEY

RAILROAD,

TO THE

PITTSBURGH, FORT WAYNE AND CHICAGO

RAILWAY COMPANY.

THIS AGREEMENT, made and concluded this twenty-ninth day of June, A. D. 1865, by and between the NEW CASTLE AND BEAVER VALLEY RAILROAD COMPANY, party of the first part, and the PITTSBURGH, FORT WAYNE AND CHICAGO RAILWAY COMPANY, party of the second part, *witnesseth:*

FIRST.—That the party of the first part, for and in consideration of the covenants and agreements of the party of the second part hereinafter mentioned, has let, leased and demised, and by these presents does let, lease and demise unto the party of the second part, all their, the party of the first part's Railroad, extending from Homewood, in Beaver County, to its northern terminus, near New Castle, in the County of Lawrence, together with side tracks, station houses, water stations, rights of way, grounds, and all appurtenances, excepting locomotives, cars, office and station furniture, tools, &c., for and during the term of ninety-nine years, commencing on the first day of July, A. D. 1865, and ending on the thirtieth day of June, A. D. 1964; together with all the right of the first party to use and operate said Railroad and appurtenances, also all their right to demand and receive fares, freight charges, tolls or other compensation, for the transportation of persons or property ; and generally their full right and authority in and over the premises, so far as the same may be necessary to enable the party of the second part to possess, enjoy and preserve said Railroad and appurtenances, agreeably to the provisions of this lease.

SECOND.—That the party of the first part has covenanted, and by these presents does covenant, that they will proceed to replace, as fast as necessary to the safety of said Railroud, the trestle-work at Stockman's and Wilson's Runs with earth embankment, carrying the water under the same by substantial stone culverts; that within twelve months they will erect, at Wampum, at Newport and Moravia, or Ziegler's, station houses and platforms sufficient for the accommodation of passengers and freight, and at the northern terminus of said Railroad, near New Castle, a freight house sufficient for the business there. That they will construct additional sidings to the extent of one mile, and at Ziegler's or at Wampum, as may best suit the convenience of the Railroad, they shall erect, and furnish with an ample supply of water, a suitable and sufficient water tank; provided that, should said party of the first part fail to perform the work above mentioned, within the time above mentioned, the party of the second part is hereby authorized to do, fairly and reasonably, all or any portion of said work, and to deduct the cost thereof from the payments due to the party of the first part, under this lease; and that, whatever other construction may be deemed necessary, in the joint judgment of the parties hereto, to accommodate the future developed business of the said Railroad, shall be made under the direction and at the cost of the party of the first part.

THIRD.—That the party of the second part, for and in consideration of the forgoing lease and covenants by the party of the first part, has covenanted, and does hereby covenant, that they will operate said Railroad, continuously, during the term of this lease, and at all times furnish to the public reasonable facilities for the transportation of persons and property, to the extent of the capacity of the track, sidings and stations; that they will at all times during said term, keep, maintain and preserve said Railroad, premises and appurtenances, and the additions and improvements that may hereafter be put thereto, in all parts thereof, in good condition and repair; that as often as any part or portion thereof, whether road-bed, rail, tie, bridge, culvert, turn-table, water-tank, station house, building, or other parcel or appurtenance whatsoever, shall, from any cause whatsoever, be destroyed, or become unfit for their appropriate uses and purposes, they, the said party of the second part, will, at their own proper cost, rebuild and renew the same; *provided*, that the same, when so rebuilt and renewed, and all other permanent structures and improvements, when established, shall become and be the property of the party of the first part, to be rebuilt, renewed and repaired by the party of the second part, as above mentioned; that they will keep and perform, for and on behalf of the party of the first part, all contracts and obligations heretofore made and entered into with other parties (a schedule whereof is hereto annexed) touching the maintaining of fences, cattle guards and passes, and will indemnify and save harmless said party of the first part from all loss or damage, by reason of second party's failure so to do; that they will indemnify and save harmless the party of the first part against and from all claims, liabilities, suits, recoveries and judgments, and from all loss and damage, whatsoever, arising or occurring in, from or by the operating said Railroad by the party of the second part.

FOURTH.—That the party of the second part has covenanted, and does hereby covenant, to pay all taxes, duties and assessments, of whatsoever name or nature, that may accrue or be assessed, charged or levied by National, State, Municipal, or other legal or competent authority, on the said Railroad, real estate and appurtenances, as the same now or hereafter may be, or upon the receipts or earnings for the transportation of persons or property, or upon the business of said Railroad, or otherwise howsoever imposed; that they will transport and pass over said Railroad of first party, free of charge, the President, Directors, Secretary and Treasurer thereof, and their respective families.

FIFTH.—That the party of the second part has covenanted, and does hereby covenant, to keep a full, true and accurate account of all business done upon said Railroad, the rates at which the same was done, and the amount of money received therefor; to permit the said party of the first part, by their authorized agent, at any time, to inspect the books of accounts and vouchers of said business; to exhibit monthly, to party of the first part, a full, particular and written statement of the gross earnings of party of the second part on the said Railroad, and within thirty days after the end of each and every month to pay to party of the first part forty per centum of the gross earnings thereof; that the forty per centum so paid, or to be paid, shall not, in the aggregate, in any year, amount to less than forty thousand dollars; that, should said per centage amount to less than forty thousand dollars, they will make good and pay to the party of the first part, within forty days after the expiration of the year, the deficiency; that is to say, such amount of money as, added to the forty per centum of earnings, would make the sum of forty thousand dollars; that the year agreed upon shall terminate on the 30th of June in each calendar year; that they will charge for business done upon said Railroad, as far as practicable, the full local rates authorized by the Act of Assembly incorporating the party of the first part, excepting and providing, nevertheless, that they may and shall keep and fulfill all contracts heretofore made with the Pennsylvania Railroad Company, the Pittsburgh, Fort Wayne and Chicago Railway Company, the Erie and Pittsburgh Railroad Company (a schedule of which is hereto annexed), for the transportation of freight and passengers.

SIXTH.—That the party of the second part has covenanted, and does hereby covenant, that in the following cases the earnings of the Railroad of first party shall be estimated and adjudged as follows, to wit: on freight from New Castle to Allegheny or Pittsburgh, or from Pittsburgh or Allegheny to New Castle, forty per centum of the whole freight charges shall be adjusted to be the earnings of said Railroad. On freight from New Castle to Rochester, or from Rochester to New Castle, sixty-two per centum; on freight from the junction of the Lawrence Railroad Company's Road with first party's Road to Allegheny or Pittsburgh, or from Allegheny or Pittsburgh to said junction, thirty-seven and one half per centum; on freight from said junction to Rochester, or from Rochester to said junction, fifty-seven per centum. On passengers over whole of first party's road, seventy cents each; on passengers from said junction to Homewood, or from Homewood to said junction, sixty cents each; *provided*, that the party of the second part may reduce the passage fare between New Castle and Allegheny or Pittsburgh, if it shall be necessary so to do, in order to compete with rival routes, in which case the earnings of first party's road shall be to the earnings on the whole route between the places aforesaid, as seventy to one hundred and eighty-five (185).

SEVENTH.—That the party of the second part has covenanted, and does hereby covenant, that if they do not pay, or cause to be paid, all and singular the taxes above stipulated to be paid by them, when thereto lawfully required, that if they fail to pay or cause to be paid to the party of the first part the per centage of the earnings upon said Rallroad, as and when the same is above stipulated to be paid, or if they fail to make good and pay the deficiency or difference between the said per centage and forty thousand dollars, as and when above stipulated to be made good and paid, and shall be in default in any or all said cases for a period of ninety days after the same should have been paid, then, in either or all of said cases of failure and default, the said party of the first part may declare and treat this lease as determined, and all rights of the second party under the same forfeited at first party's option, and thereupon may forthwith enter upon and take possession of said Railroad, premises and appurtenances, as of right, without let or hindrance of second party ; *provided*, that said party shall not thereby lose or impair their right of action or actions for the recovery of any or all debts or damages otherwise due and recoverable under the provisions of this lease, and provided, further, that in case of rightful entry by party of first part, the said party shall be entitled to recover, in addition to his debts and damages above provided for, the sum of twenty-five thousand dollars, as liquidated damages for disappointments, delays, losses and expenditures attendant and consequent upon the entry and resumption of possession.

EIGHTH.—Whereas, the said Railroad, premises and appurtenances are encumbered by a first and second mortgage, the first for one hundred and fifty thousand dollars, the second for one hundred thousand dollars, in all two hundred and fifty thousand dollars, for the non-payment of which, whether debt or interest, the said Railroad and premises, according to the conditions of said mortgages, are liable to be sold.

And whereas, it may be necessary or expedient to party of the first part to obtain an extension of time for the payment of said indebtedness, and to that end again to mortgage the Railroad and premises aforesaid, the party of the second part has covenanted, and does hereby covenant, that the first party may execute

and deliver any mortgage or mortgages in any sum or sums of
money, in the aggregate not exceeding two hundred and
fifty thousand dollars, payable at such time or times, and at
such rate of interest as by the parties thereto may be agreed
upon, and that such mortgage or mortgages, duly executed and
recorded, shall be deemed and taken as prior in date or lien to
this lease, and shall have the same force and effect, to all intents
and purposes, in law and equity, as if executed and recorded
prior to the execution of this lease, or this lease had never been
executed. *Provided, however,* that said mortgage or mortgages
shall, to the amount thereof, be in lieu and discharge of so much
of the aforesaid mortgage or mortgages now of record.

> In Testimony whereof, the Presidents of the respective
> Companies above named have hereunto set their res-
> pective hands, and the Secretaries thereof have coun-
> tersigned the same, and affixed the seals of their res-
> pective Companies, in pursuance of authority given
> them by the Board of Directors of said Companies
> respectively.

THE NEW CASTLE AND BEAVER VALLEY RAILROAD COMPANY,

Seal of the New Castle & B. V. R. R. Co.

By A. L. CRAWFORD,
President.

Attest: J. W. BLANCHARD,
Secretary.

THE PITTSBURGH, FORT WAYNE AND CHICAGO RAILWAY COMPANY,

Seal of the Pt., Ft. W. & C. Ry. Co.

By G. W. CASS,
President.

Attest: F. M. HUTCHINSON,
Secretary.

LEASE

OF

THE LAWRENCE RAILROAD TO THE PITTSBURGH,

FORT WAYNE AND CHICAGO RAILWAY

COMPANY.

THIS INDENTURE, made and entered into this twenty-second day of May, A. D. one thousand eight hundred and sixty-nine, by and between the PITTSBURGH, FORT WAYNE AND CHICAGO RAILWAY COMPANY, a corporation existing under and by virtue of the laws of Pennsylvania, Ohio, Indiana and Illinois, party of the first part, and the LAWRENCE RAILROAD COMPANY, a corporation existing under and by virtue of the laws of Pennsylvania and Ohio, party of the second part:

Whereas, the party of the first part owns and operates a Railroad, extending from Pittsburgh, in the said State of Pennsylvania, to Chicago, in the said State of Illinois, and is also the lessee of the New Castle and Beaver Valley Railroad, a line of Railroad extending from Homewood, in the said State of Pennsylvania, and situated on the said line of railroad owned by the party of the first part aforesaid, to New Castle, also in the said State of Pennsylvania; and the party of the second part owns and operates a line of Railroad extending from a point on the said New Castle and Beaver Valley Railroad, at or near Mahoningtown, in the said State of Pennsylvania, to Youngstown, in the said State of Ohio; and

Whereas, it is considered by the said parties hereto, that their mutual interests will be promoted by having the said

Railroads, forming a continuous line between Pittsburgh and Youngstown, *via* Homewood and Mahoningtown aforesaid, placed under one management and control, upon the terms and conditions hereinafter expressed:

Now, THEREFORE, THIS INDENTURE WITNESSETH, that the said parties of the first and second part, their successors and assigns, in consideration of the premises, and for and in the further consideration of one dollar in hand paid to the said party of the second part by the said party of the first part, at or before the ensealing and delivery hereof, the receipt of which is hereby acknowledged, have covenanted, promised and agreed, and by these presents do covenant, promise and agree, to and with each other, for themselves and their successors and assigns, in manner and form following, that is to say:

First.—The said party of the second part, for itself and its successors and assigns, has let, leased and demised, and by these presents does let, lease and demise unto the said party of the first part, its successors and assigns, for and during the term of ninety-nine years, commencing with the date hereof, the Railroad of the said party of the second part, extending from a point on the New Castle and Beaver Valley Railroad, at or near Mahoningtown, to Youngstown, as aforesaid, together with all and singular the side tracks, station house, water stations, machine shops, engine houses, turn-tables and other buildings, lands, rights of way and all other appurtenances, in any manner thereunto belonging, together with all the right of the said party of the second part to use and operate the said railroad and appurtenances; also the right of the said party of the second part to demand and receive fares, freight charges, tolls, or any other compensation for the transportation of persons or property; and generally the full right and authority of the said party of the second part, in and over the said railroad and its appurtenances, so far as may be necessary to enable the said party of the first part, its successors and assigns, to fully possess, enjoy, and preserve the said railroad and its appurtenances, agreeably to the provisions of this lease.

Second.—The said party of the second part shall proceed with, as fast as practicable, and finish the fencing of the said railroad, necessary to be done under the laws of Ohio, and construct such additional cattle guards as may be, in the joint judgment of the parties hereto, necessary for the safe working of the said railroad, and shall also complete the reservoir at or near Hillsville Station, with a capacity sufficient for the proper supply of water for the locomotives in use on the said railroad, and generally such additional side tracks, station buildings, and whatever other construction may be deemed necessary, in the joint judgment of the parties hereto, to accommodate the future developed business of the said railroad.

Third.—The said party of the second part shall protect, save harmless, and indemnify the said party of the first part, its successors and assigns, from and against all claims, demands or suits for right of way, and for any and all injury to property, arising out of, or appertaining to, or connected with the construction or building of the railroad aforesaid.

Fourth.—The said party of the first part, its successors and assigns, may and shall continuously operate during the term of this Indenture, the said railroad and its appurtenances hereby leased and demised, and shall at all times furnish to the public all reasonable facilities for the transportation of persons and property, to the extent of the capacity of the track, side tracks, and stations, and other buildings, and shall at all times, during the said term, keep, maintain and preserve the said railroad and appurtenances, and the additions and improvements that may hereafter be put thereto, in all parts in good condition and repair ; that as often as any part or portion of the said railroad, or any of its appurtenances, shall from any cause be destroyed or otherwise become unfit for their appropriate uses and purposes, the said party of the first part, and its successors and assigns, shall, at its own cost and expense, renew or rebuild the same, which said renewed structures shall at once become the property of the said party of the second part, and the said party of the first part, its successors and assigns, shall and will, from time to time, furnish for use upon the said demised railway any and all rolling stock and equipment which the business of the said demised railroad, and the increase thereof from time to time, may require.

Fifth.—The said party of the first part, its successors and assigns, shall pay all taxes, duties and assessments, whatsoever, that may be assessed, charged or levied by national, state, municipal or other legal authority, on the said real estate and appurtenances hereby leased and demised, and upon the earnings or receipts for the transportation of persons or property over the said railroad, or otherwise upon the business of the said railroad, *provided* that nothing in this instrument contained shall be so construed as to render the said party of the first part, or its successors and assigns, liable for the tax specifically upon the interest on the bonds, nor upon the dividends on the stock of the said party of the second part.

Sixth.—The said party of the first part shall keep a full, true and accurate account of all business done upon the said railroad, during the existence of this lease, the rates at which the same was done, and the money received therefor, and shall permit the said party of the second part, by its duly authorized agent, at any time, to inspect the books of account and vouchers of said business, and shall furnish monthly to the said party of the second part a full and accurate written statement of the gross earnings of the said party of the first part on the railroad hereby leased and demised, and shall, within forty-five days after the end of each calendar month, pay to the said party of the second part forty per cent. of such gross earnings ; but it is hereby expressly agreed that the aforesaid payments shall amount, in each and every year, to at least forty-five thousand dollars, which is hereby agreed upon as a *minimum* amount to be paid in each and every year, and which is to be paid absolutely without reference to the per centage which it forms of the gross earnings of such year, and without leaving or creating any charge upon the earnings of any future year; it being also understood and agreed that each year under this Indenture shall commence on the first day of June and terminate on the thirty-first day of May in each and every calendar year ; and the said party of the first part, its successors and assigns, shall be entitled to retain in each and every year of the term aforesaid all excess of gross earnings for such year over and above the payments to the said party of the second part above provided, and to apply the same to and for the purposes of this Indenture and for fulfilling all the undertakings of the said party of the first part herein expressed, and to apply to its own use and benefit any surplus that may remain in such year, as compensation for the services, acts and things done, or to be done, by the said party of the first part, its successors and assigns, in pursuance of these presents.

Seventh.—The said party of the first part, its successors and assigns, shall charge for the business done on the said railroad, so far as practicable, the full local rates authorized by the charter of the said party of the second part, but the said party of the first part, its successors and assigns, may, when necessary, make such special rates as will secure competitive business, and fully develop otherwise the business of the said railroad, it being thereby expressly understood and agreed, that when persons or property are carried on through tickets, or through bills of lading or manifests, from points on the said railroad to points on other railroads, or *vice versa*, the earnings therefrom allotted to the said railroad, shall be *pro rata* on the basis of the number of miles which such persons or property may be so transported.

Eighth.—In case the said party of the first part, its successors and assigns, shall make default in the payment of any taxes hereinbefore stipulated to be paid by it, after being lawfully required so to do, or if the said party of the first part, its successors and assigns, shall make default in the payment of the fixed *minimum* aforesaid, agreed to be paid under this Indenture, or of the forty per cent. of the gross earnings aforesaid, in the manner and at the times hereinbefore provided, to be paid to the said party of the second part, and such default shall continue for the period of ninety days, the said party of the second part may, at its election, annul and terminate this lease, by written notice, to the said party of the first part, or its successors and assigns, and thereupon may forthwith enter upon and take possession of the said railroad, real estate and appurtenances, as of right, and without let or hindrance, by the said party of the first part, or its successors and assigns, or it may take such other and further action for the enforcement of the provisions of this Indenture as to it may seem advisable; *provided*, that by the taking of such possession the said party of the second part shall not lose or impair its right of action at law for the recovery of any and all debts or damages otherwise due and recoverable under the provisions of this Indenture ; and *provided*, further, that in case of rightful entry by the said party of the second part, it shall be entitled to recover, in addition to its debts and damages above provided for, the sum of twenty-five thousand dollars, as liquidated damages for disap-

pointments, delays, losses and expenditures attendant and consequent upon such entry and resumption of possession.

Ninth.—The said party of the first part, its successors and assigns, shall, at all times during the term aforesaid, bear, and at
its own proper cost and expense, pay and discharge any and all
costs, expenses and charges whatsoever, of operating and transacting the business and maintenance of the railroad aforesaid,
or in any manner connected with, arising out of, or appertaining
to the business, operation, maintenance or management of the
same, and shall at all times, during the term aforesaid, hold, save
and keep harmless and indemnify the said party of the second
part, from and against any and all charges, costs and expenses,
suits, damages and claims of any and all kinds whatsoever, arising out of or in any manner appertaining to or connected with
the operation, maintenance or management, during the term
aforesaid, of the said railroad and appurtenances, including not
only the expenses of operating, maintaining and managing the
said railroad and appurtenances, but, also, any and all claims
for injury to persons or property that may occur upon the said
railroad during the term aforesaid, and also any and all claims,
demands or suits for non-performance, or breach of contract, in
respect to any person or thing to be transported over the same,
and also any and all claims, demands or suits for the loss or destruction, by whatever cause, of any property whatsoever, while
under the control of the said party of the first part, its successors and assigns, or which it shall have undertaken to carry or
transport over any portion of the said railroad.

Tenth.—The said party of the first part shall and will, at all
times, during the term hereby granted, send over the said
demised railway any and all persons and property to be carried
or transported by it or them from Pittsburgh, or from any
point on the line of the railway of the party of the first part, or
of the Beaver Valley Railway, lying between Pittsburgh and
New Castle, or from any point westerly of Homewood, to any
point or place situate upon or in the direction of the line of the
said demised Lawrence Railroad, and also any and all persons
and property to be carried or transported by the said party of
the first part, its successors or assigns, from Youngstown or any
point beyond, or from any point between Youngstown and

Mahoningtown, to any point or place situate upon or in the direction of the line of the said demised railroad; and the said party of the first part, for itself, its successors and assigns, further covenants, promises and agrees to and with the party of the second part, that the said party of the first part, its successors or assigns, shall not, and it or they will not at any time during the term hereby granted, construct or aid in the construction of, or lease, operate or make running connections with any line of railroad which shall be parallel to or run in the same general direction with the said demised railroad, and which shall reach or connect with the Beaver Valley Railroad, or the Pittsburgh, Fort Wayne and Chicago Railway, at any point or place situate between New Castle and Pittsburgh.

Eleventh.— *Whereas*, the said railroad, real estate and appurtenances are encumbered by a first and only mortgage to secure the payment of three hundred and sixty thousand dollars of bonds, bearing date the first day of August, A.D. one thousand eight hundred and sixty-five, and due and payable on the first bay of August, A.D. one thousand eight hundred and ninety-five, for the non-payment of which bonds, or of the accrued interest thereon, the said railroad, real estate and appurtenances are liable to be sold under a foreclosure of such mortgage; and

Whereas, it may be necessary or expedient for the said party of the second part to obtain an extension of time for the payment of such bonds, or to substitute new bonds therefor at the maturity thereof, and to that end again to mortgage the said railroad, real estate and appurtenances, it is hereby covenanted and agreed by and between the parties hereto, for themselves, their successors and assigns, that the said party of the second part may extend such bonds, or substitute others-therefor, as may be agreed upon by the parties thereto, secured by the present or new mortgage or deed of trust, in the same manner and with the like effect as if these presents had not been executed, but the aggregate amount of such bonds, or of the annual interest charge created thereby, shall not be increased, and in default of such extension or substitution, the said party of the first part, its successors and assigns, shall have the right to apply to the payment of such overdue bonds, all the surplus over the interest on the bonds outstanding, from the payments to which the said party of the second part would be otherwise entitled; and in case the said party of the second part shall at any time or times make default in the payment of any interest on the bonds aforesaid, which shall have been duly demanded, on any day on which the same shall be payable, or shall make default in the payment of any instalment of the sinking fund therefor, on any day when the same shall become payable, the said party of the first part, its successors and assigns, may, without waiting for the expiration of the three months before such default can be availed of by the bondholder, proceed at once to pay such interest, and to pay any instalment of the Sinking Fund that may be overdue, and to deduct the amount of the payments therefor out of the next succeeding payment for account of the *minimum* aforesaid, or of the forty per cent. of gross earnings aforesaid, with interest at the rate of seven per cent. from the time of such advances until the time when such payment would fall due, and every such advance, with interest as aforesaid, shall be duly credited to the said party of the first part, its successors and assigns, in the annual settlement between the parties hereto.

IN WITNESS WHEREOF, the said parties hereto have caused their respective corporate seals to be hereto attached, and the same to be attested by the signatures of their respective Presidents and Secretaries, the day and year first above written.

PITTSBURGH, FORT WAYNE AND CHICAGO RAILWAY CO.,

By G. W. CASS,
President.

Seal
P., F. W. & C.
Railway Co.

F. M. HUTCHINSON,
Secretary.

Sealed and delivered in
presence of—
W. R. SHELBY,
JAMES H. THAW.

LAWRENCE RAILROAD COMPANY,

By WM. McCREARY,

President.

```
. . . . . . . . . . . .
:      Seal       :
:    Lawrence     :
:    R. R. Co.    :
. . . . . . . . . . . .
```

F. M. HUTCHINSON,

Secretary.

In presence of—

JAMES H. THAW,

JNO. H. FREDRICK.

LEASE

OF

THE MASSILLON AND CLEVELAND RAILROAD

OF THE

PITTSBURGH, FORT WAYNE & CHICAGO RAILWAY COMPANY.

THIS INDENTURE, made and entered into this twenty-second day of May, A. D. one thousand eight hundred and sixty-nine, by and between the PITTSBURGH, FORT WAYNE AND CHICAGO RAILWAY COMPANY, a corporation existing under and by virtue of the laws of Pennsylvania, Ohio, Indiana and Illinois, party of the first part, and the MASSILLON AND CLEVELAND RAILROAD COMPANY, a corporation existing under and by virtue of the laws of Ohio, party of the second part:

Whereas, the party of the first part owns and operates a Railroad, extending from Pittsburgh, in the said State of Pennsylvania, to Chicago, in the said State of Illinois, and is also the owner of the Cleveland, Zanesville and Cincinnati Railroad, a line of Railroad extending from a point four miles south of Millensburg to Hudson, on the Cleveland and Pittsburgh Railroad, in the said State of Ohio, and forming a line, *via* the said Cleveland and Pittsburgh Railroad, to Cleveland, Ohio, party of the second part, is engaged in building a Railroad, extending from a point on the Railroad of the said party of the first part, at or near Massillon, to Clinton on the said Cleveland, Zanesville and Cincinnati Railroad, in the said State of Ohio, and,

Whereas, it is considered by the said parties hereto, that their mutual interests will be promoted by having the said Railroads, forming a continuous line between Massillon and Hudson, *via* Clinton aforesaid, placed under one management and control, upon the terms and conditions hereinafter expressed.

Now, THEREFORE, THIS INDENTURE WITNESSETH, That the said parties of the first and second part, their successors and assignes, in consideration of the premises, and for and in the further consideration of one dollar in hand paid to the said party of the second part, by the said party of the first part, at or before the ensealing and delivery hereof, the receipt of which is hereby acknowledged, have covenanted, promised and agreed, and by these presents do covenant, promise and agree, to and with each other, for themselves and their successors and assigns, in manner and form following, that is to say :

First.—The said party of the second part, for itself and its successors and assigns, has let, leased and demised, and by these presents does let, lease and demise unto the said party of the first part, its successors and assigns, for and during the term of ninety-nine years, commencing with the date of the completion, and its delivery to the said party of the first part, of the Railroad of the said party of the second part, extending from a point on the Railroad of the said party of the first part, at or near Massillon, to Clinton, on the said Cleveland, Zanesville and Cincinnati Railroad as aforesaid ; together with all and singular the side tracks, station houses, water stations, machine shops, engine houses, turn tables, and other buildings, lands, rights of way, and all other appurtenances, in any manner thereunto belonging, together with all the right of the said party of the second part to use and operate the said Railroad and appurtenances ; also the right of the said party of the second part to demand and receive fares, freight charges, tolls, or any other compensation, for the transportation of persons or property, and generally the full right and authority of the said party of the second part in and over the said Railroad and its appurtenances, so far as may be necessary to enable the said party of the first part, its successors and assigns, to fully possess, enjoy and preserve the said Railroad and its appurtenances agreeably to the provisions of this lease.

Second.—The said party of the second part shall proceed with, as fast as practicable, and finish, the building of the said Railroad, together with the fencing necessary to be done under the laws of Ohio, and construct such cattle guards as may be, in the joint judgment of the parties hereto, necessary for the safe working of the said Railroad, and shall also construct such side tracks, station buildings, water houses and appurtenances, as may be necessary for the proper and economical working of the said Railroad, and the suitable accommodation of the business to be done on the said Railroad, and generally such additional side tracks, station buildings and whatever other construction may be deemed necessary, in the joint judgment of the parties hereto, to accommodate the future developed business of the said Railroad.

Third.—The said party of the second part shall protect, save harmless and indemnify the said party of the first part, its successors and assigns, from and against all claims, demands, or suits for right of way, and for any and all injury to property arising out of or appertaining to or connected with the construction or building of the railroad aforesaid.

Fourth.—The said party of the first part, its successors and assigns, may and shall contiuously operate, during the term of this Indenture, the said Railroad, and its appurtenances, hereby leased and demised, and shall at all times furnish to the public all reasonable facilities for the transportation of persons and property, to the extent of the capacity of the track, side tracks, and stations, and other buildings, and shall at all times, during the said term, keep, maintain, and preserve the said Railroad and appurtenances, and the additions and improvements that may hereafter be put thereto, in all parts, in good condition and repair; that as often as any part or portion of the said Railroad, or any of its appurtenances, shall from any cause be destroyed, or otherwise become unfit for their appropriate uses and purposes, the said party of the first part, and its successors and assigns, shall, at its own cost and expense, renew or rebuild the same—which said renewed structures shall at once become the property of the said party of the second part.

Fifth.—The said party of the first part, its successors and assigns, shall pay all taxes, duties and assessments, whatsoever, that may be assessed, charged or levied, by national, state, municipal, or other legal authority, on the said railroad, real estate and appurtenances hereby leased and demised, and upon the earnings or receipts for the transportation of persons or property over the said Railroad, or otherwise, upon the business of the said Railroad; *provided*, that nothing in this instrument contained shall be so construed as to render the said party of the first part, or its successors and assigns, liable for the tax specifically upon the interest on the bonds, nor upon the dividends on the stock of the said party of the second part.

Sixth.—The said party of the first part shall keep a full, true and accurate account of all business done upon the said Railroads, during the existence of this lease, the rates at which same was done, and the money received therefor, and shall per. mit the said party of the second part, by its duly authorized agent, at any time to inspect the books of account and vouchers of said business, and shall furnish, monthly, to the said party of the second part a full and accurate written statement of the gross earnings of the said party of the first part on the Railroad hereby leased and demised, and shall, within forty-five days after the end of each calendar month, pay to the said party of the second part forty per cent. of such gross earnings; but it is hereby expressly agreed that the aforesaid payments shall amount, in each and every year, to at least twenty thousand dollars, which is hereby agreed upon as a *minimum* amount to be paid in each and every year, and which is to be paid absolutely, without reference to the per centage which it forms of the gross earnings of such year, and without leaving or creating any charge upon the earnings of any future year—it being also understood and agreed that each year, under this Indenture, after this present year, shall commence on the first day of January, and terminate on the thirty-first day of December, in each and every calendar year; and the said party of the first part, its successors and assigns, shall be entitled to retain, in each and every such year of the term aforesaid, all excess of gross earnings for such year, over and above the payments to the said party of the second part above provided, and to apply the same to and for the purpose of this Indenture, and for fulfilling all the undertakings of the said party of the first part herein expressed, and to apply to its own use and benefit any surplus, that may remain in such year, as compensation for the services, acts and things done, or to be done, by the said party of the first part, its successors and assigns, in pursuance of these presents—it being expressly understood and agreed, that the present year, by reason of the said Railroad being yet unfinished, shall be deemed and taken as a fractional year, and the payments and other things to be done by the said party of the first part, under this lease, shall date or take effect from the completion and de-livery, by the said party of the second part, of the said Railroad, to the said party of the first part.

Seventh.—The said party of the first part, its successors and assigns, shall charge for the business done on the said Railroad, so far as practicable, the full local rates authorized by the charter of the said party of the second part, but the said party of the first part, its successors and assigns, may, when necessary, make such special rates as will secure competitive business, and fully develop otherwise the business of the said Railroad, it being hereby expressly understood and agreed that when persons or property are carried on through tickets, or through bills of lading or manifests from points on the said Railroad to points on other Railroads, or, *vice versa,* the earnings therefrom allotted to the said railroad shall be *pro rata* on the basis of the number of miles which such persons or property may be so transported; and it is also hereby expressly understood and agreed that all coal offered for shipment over the Railroad hereby leased and de-mised, and destined for Cleveland or other lake ports, shall always be forwarded *via* Hudson and the Cleveland and Pitts-burgh Railroad aforesaid, so long, and whenever, the same can be done on terms equally favorable with other routes; and no terms for such transportation shall be granted by the party of the first part, its successors and assigns, *via* any other route to such lake ports, than are granted to the line or route *via* Hudson and the Cleveland and Pittsburgh Railroad aforesaid.

Eighth.—In case the said party of the first part, its succes-sors and assigns, shall make default in the payment of any taxes, hereinbefore stipulated to be paid by it, after being lawfully re-quired so to do, or if the said party of the first part, its succes-sors and assigns, shall make default in the payment of the fixed *minimum* aforesaid, agreed to be paid under this Indenture, or of the forty per cent. of the gross earnings aforesaid, in the manner and at the times hereinbefore provided, to be paid to the said party of the second part, and such default shall continue for the period of ninety days, the said party of the second part may, at its election, annul and terminate this lease, by written notice to the said party of the first part, or its successors and assigns, and thereupon may forthwith enter upon and take possession of the said Railroad, real estate and appurtenances, as of right and without let or hindrance by the said party of the first part, or its successors and assigns ; or it may take such other and further action, for the enforcement of the provisions of this Indenture, as to it may seem advisable : *provided*, that by the taking of such possession, the said party of the second part shall not lose, or impair, its right of action, at law, for the recovery of any and all debts or damages otherwise due and recoverable under the provisions of this Indenture ; and, *provided, further*, that in case of rightful entry by the said party of the second part, it shall be entitled to recover, in addition to its debts and damages above provided for, the sum of fifteen thousand dollars, as liquidated damages for disappointments, delays, losses and expenditures attendant and consequent upon such entry and resumption of possession.

Ninth.—The said party of the first part, its successors and assigns, shall at all times, during the term aforesaid, bear, and at its own proper cost and expense, pay and discharge any and all costs, expenses and charges, whatsoever, of operating and transacting the business and maintenance of the Railroad aforesaid, or in any manner connected with, arising out of, or appertaining to the business, operation, maintenance or management of the same ; and shall, at all times during the term aforesaid, hold, save and keep harmless and ,indemnify the said party of the second part from and against any and all charges, costs and expenses, suits, damages and claims, of any and all kinds whatsoever, arising out of or in any manner appertaining to or connected with the operation, maintenance or management, during the term aforesaid, of the said Railroad and appurtenances, including not only the expenses of operating, maintaining and managing the said Railroad and appurtenances, but also any and all claims for injury to persons or property that may occur upon the said Railroad during the term aforesaid, and also any and all claims, demands or suits for non-performance or breach of contract in respect to any person or thing to be transported over the same, and also any and all claims, demands or suits for the loss or destruction, by whatever cause, of any property whatsoever, while under the control of the said party of the first part, its successors and assigns, or which it shall have undertaken to carry or transport over any portion of the said Railroad.

Tenth.— Whereas, it may be necessary, or expedient, for the said partyof the second part, for the purpose of realizing the means necessary to fully complete its said Railroad and appurtenances, to create and issue its bonds, and to secure the payment of the same, at their maturity, by a mortgage or deed of trust, conveying the said Railroad, appurtenances and franchises, in trust for the benefit of the holders of the said bonds, it is hereby covenanted and agreed by and between the parties hereto, for themselves, their successors and assigns, that the said party of the second part may create and issue such bonds, in such manner and form, and bearing such a rate of interest, not exceeding seven per cent. per annum, as may by them be deemed proper and expedient, and may secure the payment of the said bonds by a mortgage or deed of trust, conveying the said Railroad, appurtenances and franchises in trust for the benefit of the holders of the said bonds, in the same manner and with like effect as if these presents had never been executed ; and may also create a sinking fund for the redemption of the said bonds: *provided,* and it is hereby expressly declared and agreed, by and between the parties hereto, their successors and assigns, that in no event shall the amount of the said bonds exceed the sum of one hundred thousand dollars, and the said mortgage, duly executed and recorded, shall be deemed and taken as prior in date and lien to this Indenture, and shall have the same force and effect, to all intents and purposes, in law and in equity, as if it had been executed and recorded prior to the execution of this Indenture, or if this Indenture had never been executed; and the said party of the second part may, if it shall become necessary or expedient, extend or renew the said bonds, or substitute others therefor, as may be agreed upon by the parties thereto, secured by the then existing or a new mortgage or deed of trust, in the same manner and with like effect as if these presents had not been executed, but the aggregate amount of such bonds, or of the annual interest charge created thereby shall not be increased ; and in default of such extension or substitution, the said party of the first part, its successors and assigns, shall have the right to apply, to the payment of such overdue bonds, all the surplus over the interest on the bonds outstanding, from the payments to which the said party of the second part would be otherwise entitled ; and in case the said party of the second part shall, at any time or times, make default in the payment of any interest on the bonds aforesaid, which shall have been duly demanded, on any day on which the same shall be payable, or shall make default in the payment of any instalment of the Sinking Fund therefor, on any day when the same shall become payable, the said party of the first part, its successors and assigns, may, without waiting for the expiration of the three months before such default can be availed of by the bondholders, proceed at once to pay such interest, and to pay any instalment of the Sinking Fund that may be over due, and to deduct the amount of the payments therefor out of the next succeeding payment for account of the *minimum* aforesaid, or of the forty per cent. of gross earnings aforesaid, with interest, at the rate of seven per cent., from the time of such advances until the time when such payments would fall due, and every such advance, with interest as aforesaid, shall be duly credited to the said party of the first part, its successors and assigns, in the annual setttlement between the parties hereto.

In WITNESS WHEREOF, the said parties hereto have caused their respective corporate seals to be hereto affixed, and the same to be attested by the signatures of their respective Presidents and Secretaries, the day and year first above written.

PITTSBURGH, FORT WAYNE AND CHICAGO RAIL-WAY CO.,

By G. W. CASS,
President.

Seal. P., F. W. & C. Ry. Co.

Attest—

F. M. HUTCHINSON,
Secretary.

Sealed and delivered
in the presence of—

W. R. SHELBY,
J. N. McCULLOUGH.

MASSILLON. AND CLEVELAND RAILROAD CO.,

By SIMON PERKINS,

President.

Attest—

```
...........
:   Seal   :
: M. & C. R. R. :
:    Co.    :
...........
```

F. M. HUTCHINSON,

Secretary.

Sealed and delivered
in the presence of—

JAMES H. THAW,
JNO. H. FREDRICK.

CIRCULAR

TO THE

STOCKHOLDERS AND BONDHOLDERS

OF THE

PITTSBURGH, FORT WAYNE & CHICAGO

RAILWAY COMPANY.

During the last three years you have been advised that the relations between this Company and the Pennsylvania Railroad Company have engaged the earnest attention of the President and Directors of this Company, and have been in several instances the subject of laborious and protracted negotiations between the representatives of the two Companies.

The source of the complications which tended, on occasions often recurring, to disturb the harmony of these relations, was in facts these:

At Pittsburgh, your only eastern connection with New York and the other points of your principal business to and from the east, is the Pennsylvania Railroad. Not only had the Company owning that Railroad another western connection, but it had gradually acquired a vast and permanent pecuniary interest in promoting that connection at your expense. As owner of a controlling majority of the stock of the Pittsburgh, Columbus and Cincinnati Railroad, and guarantor of the recent lease by that Company of the Columbus, Chicago and Indiana Central Railway, it has become the practical proprietor of a line to Cincinnati superior to your own, of an equal line to St. Louis, and of an inferior line to Chicago.

Such a situation between these two Companies, possessing the shortest, most cheaply operated, and most effective line between New York and Chicago, and capable of serving the public on the most economical terms, and yet affording reasonable returns to their proprietors, was intrinsically false, and was liable to convert natural allies into rivals and foes.

The position, if it did not generate injustice, could not but create suspicion of it; and the executive men of the two Companies were exposed to the influence of a consciousness of discordant interests and the attrition of daily rivalry.

No matter how the situation came to exist, it was to be dealt with as a fact. What an ealier foresight or an ealier ability might have avoided, was now, if possible, to be remedied.

Two or three years ago there was a negotiation on the basis that the Pennsylvania Railroad Company should retire from all interests west of Pittsburgh; but the plan was full of difficulties, and failed.

Last year, in June and in December, at prolonged conferences between Committees of the respective Boards, three different plans were discussed. Consolidation, lease, and a permanent contract, aiming to regulate the relations of the parties were considered. Nothing but the latter seemed to be capable of being agreed upon. The basis of such a contract was adopted, but the complexity of the relations to be regulated through future years, and the difficulty of finding adequate legal methods of enforcing, the necessary stipulations, rendered such a contract less secure and reliable than either of the other plans.

While this arrangement was being perfected, an unexpected event happened which changed the character of the negotiation. In January, the Pennsylvania Railroad Company felt compelled to protect its connections with St. Louis and other western points by a permanent lease of the Columbus, Chicago and Indiana Central Railway, in the name of a Company a majority of the stock of which it owns and under its own guaranty.

Both parties recognize in this event a new occasion and an increased motive to renew the effort to place their relations upon a permanently harmonious and safe foundation. The result of the conferences between the Committees of the respective Boards was an agreement on the general basis of a lease; and the last six weeks have been busily employed by the Committees and officers of the two Companies, with two concurrent meetings of the respective Boards, in perfecting the details of a lease, which was executed by the Pennsylvania Railroad Company on the seventh day of the present month.

During these years of negotiation, the idea of forming an independent connection from Pittsburgh to the East has been fully considered, and an excellent route examined. That measure would have involved the assumption of large liabilities at the outset—the construction of a costly link—a conflict with this powerful corporation before our line could be ready for business—and the creation of two complete lines from the seaboard to the lakes; while the combination of the present lines could be made to furnish the best possible route, and their capacity to accommodate a vast increase of business, and at less rates than the new lines could afford, would be secured at a comparatively small outlay.

In the meantime, the policy of the Company has been to strengthen itself by re-investing its surplus earnings in the improvement of its line, and enlarging its capacity, both as to the amount and the economy of its business.

And now that terms have been obtained which are reasonably satisfactory to our stockholders—which enhance the security of our bondholders—and which are permanent and reliable—we cannot hesitate to accept the Pennsylvania Railroad Company as an ally instead of a rival; and although, perhaps, the income of our stockholders and bondholders might be deemed already certain, we cannot doubt that they will prefer, instead of rivalry and perhaps conflict, the superadded guaranty of a wealthy corporation, adding to our annual two millions over interest, its annual four millions and a half over interest, for our security.

There is no greater error than the narrow and selfish idea that one party necessarily loses what another party gains. On the contrary, the true basis of all exchanges is that they are mutually beneficial. And the just and safe foundation for all contracts like that now proposed is that they are so framed as to promote the permanent interests of both parties.

We believe that the Pennsylvania Railroad Company will not only assure itself without cost a valuable connection, and the resulting profits on its own line, for which sacrifices are often made—but that it will derive a liberal profit from our line.

Our bondholders—of the first and second classes of mortgages —whose securities already range higher than any similar ones in our market, are still capable of being benefitted. Experience shows that bonds of the first class in pecuniary and legal security participate more or less in any depreciation which affects the stock of the Company by which they are issued, whether it results from uncertain or hazardous undertakings, or from corporate mismanagement or discredit. And experience also shows that the limit has scarcely been found of the premium which some classes of investors will pay for enhanced safety in their incomes.

The income bonds were originally subject to the contingency of a failure of net earnings to meet the interest on them. The prosperity of the Company—the vast outlay which has been made upon its line, equal to about half its present cost, as represented on the books—has rendered that contingency merely nominal. But, by the present contract, this contingency is removed, and the right to receive the interest on these bonds from the rental becomes absolute.

The stockholders become entitled to the benefit of a fund adequate to give them twelve per cent. on the present amount of stock, free and clear of all taxes, whether Federal or State, which are collected in any manner through the corporation, and are guaranteed perpetually in the enjoyment of that fund.

In this connection it is proper to advert to the proposal to convert the present stock, which would be entitled to dividends of twelve per cent., free of deductions for taxes, into a new stock, which would be entitled to dividends of seven per cent., quarterly, free of such deductions.

That measure is, in this instance, totally clear from every question of personal interest or public policy, except whether it will be advantageous to the stockholders. The annual sum, receivable in quarterly instalments, is the same in either case, and the security to which that annual sum furnishes the income is perpetual.

The proposed new capitalization has no element in common with what is sometimes called the *watering* of stocks. It involves none of the expectation of equal dividends on increased issues, which lends attraction, sometimes illusory, to such financial expedients. It does not capitalize temporary prosperity, favorable seasons or abundant harvests, or speculative anticipations of the future, or the sanguineness of human hopes. It

capitalizes only a fixed and perpetual money income, better secured than almost any First Mortgage Bond.

The State of Ohio indicated her public policy in providing
for àn annual remuneration to the investor in railroad stocks of
ten per cent. from the time of the investment, and a reasonable
probability of the continuance of that income, as a condition
which should precede and regulate her exercise of the power to
reduce charges in favor of the public.

Measured by that standard, the twenty per cent. or $2,300,000
of the increased capital would exceed that limit, if the investors
had in the past received the remuneration contemplated, and 51¾
per cent. would be a mere change of the form of the security.

But in fact the nominal par of the stock is far below the
amount of actual money invested, and far below the actual cost
to the proprietors.

The original Company which owned the entire line was formed
by a consolidation of three corporations, which took effect August
1, 1856. The capital issued was about $6,250,000. On the
reorganization under the present Company, which took effect
May 1, 1862, the authorized capital was $6,500,000 ; but in the ad-
justments, about $250,000 was saved and sold, and the proceeds
applied to construction.

The failure of the original Company resulted exclusively from
the attempt to complete the line by temporary loans, and the
inability to continue that system during and after the revulsion
of 1857.

In January, 1860, when the property went into the possession
of a receiver, at the instance of the bondholders, represented in
part by the undersigned, the stock had been the three and a half
years after the consolidation without dividend ; and, averaging
the investments before that period, it probably had been more
than five years without dividend. From that time until January, 1864—four years longer—no dividend was paid ; the first
quarterly dividend of 2½ per cent. being paid for the quarter
ending March 31st, 1864, under the new arrangement unanimously agreed to by the bondholders and stockholders in their con-
vention of that year.

During that four years, about three and three quarter millions
of dollars of net earnings were re-invested in improving and
equipping the line, and all of it was invested economically and
productively. In all, nearly seven millions of net earnings have

been thus re-invested, without counting interest thereon, and without adding a dollar to the stock or bonds of the Company; and detailed statements of the nature of the investments have been annually made; and the property which has been added remains visibly to attest the wisdom and prudence of the policy. The agreement for re-organization between the bondholders and stockholders was made in view of the railway revulsion of 1857; was conceived in a spirit of caution and sobriety; all classes made their contributions to place the property upon a sound and durable basis; the bondholders conceded about a million from their legal rights, and more than half a million was saved by the prudent and skillful administration of the trust created by the creditors after the creditors were paid. These results, with abstinence from dividends, and surplus earnings, have supplied the means whereby seven millions in cash cost, and ten or twelve millions in present value, of improvements, have been added to the property. The real estate of the Company at Chicago, Allegheny City and Pittsburgh, is represented on the books at not more than from one-third to one-half its original cost, not more than a tenth of its present value. The five millions of new stock issued since 1864 were all issued for cash at par, except $200,000 of concession made in the price of one issue of a million, which was allotted *pro rata* to the stockholders during a period of depression.

If stock had been issued for the money which has been invested but not represented, and for dividends thereon, at the rate of seven per cent., or, if stock had been issued for dividends withheld, the present amount of the stock would have been doubled. If ten per cent., instead of seven, had been adopted as the proper rate of remuneration, and stock issued accordingly, the present amount of the stock would have been much more than doubled.

The conservative policy thus far followed was induced by a sense of the ever-recurring wants incident to so rapid a development of the business—which has quadrupled, at the same rate of charges and on the same line, in the nine years of the present administration—by a belief that a system so transcendently successful had not completely matured its proper fruits; by a desire to strengthen and solidify the finances of the Company, so as to assure the maintenance of dividends at the rate begun in 1864, and to enable it to surmount adverse changes possible to happen in the general business of the country, or in its own relation; and by a just repugnance to begin what would not be certain to continue, or to excite expectations in the public, or in innocent investors, which might not be fully realized. That conservative policy has been adhered to against repeated combinations to obtain control of the Company in order to reverse it for purposes of temporary speculation; and it would probably have been overthrown but for the last vote of the bondholders; and might perhaps have at last succumbed.

Now that the Company is relieved from providing for the future wants of the business, and from all contingencies affecting that business, these motives cease, and there seems to be no great objection to the stockholders being allowed to capitalize their fixed and perpetual money income in such manner as they please.

A seven per cent. stock, on which dividends would be guaranteed by net earnings of five times their annual amount, may be considered a more certain and reliable security than almost any first mortgage bond of the best class, while it has many peculiar advantages for the investor.

1. It is perpetual, which is a valuable quality, if we assume that the rate of interest has a tendency to fall as capital increases and population becomes dense.

2. It is safe from robbery, theft, loss, or destruction by fire. The certificate is a mere evidence of the rights of the holder of the stock; and no casualty to it would impair these rights or transfer them to another party. Bonds have the character which the law merchant has by a special and exceptional rule imparted to negotiable paper. Any innocent person who obtains a railroad bond or coupon for value acquires a superior title to the original holder, even though he may have lost possession by burglary, theft, or accident. In the frequency of such occurrences of late years, in the care attending the custody of such a delicate kind of property, in the inconveniences and hazard in cases where the owners are absent for long periods, or are minors or incompetent persons—the superiority of this form of investment is very important.

3. It is more convenient to large classes of investors to collect a dividend in a single sum, than to cut off and present numerous coupons.

4. It is more productive of income. Bonds are subject to

the deduction of five per cent. of the United States income tax, which is collected through the Company liable to pay the interest, and made by law the agent of the Government to collect the income tax. They are liable to similar taxes which may be imposed by the States, as has been done by Pennsylvania. All such taxes, and all others which are collected through the corporation owning the property, or the corporation leasing it, or by virtue of jurisdiction over the property itself, are to be paid by the lessees. In another respect there is an important difference. Bonds are usually deemed to follow the person of the holder, and to be taxable wherever he resides. Stocks are mere representatives of interests in the ownership of the railway, which consists nearly all of real property that is local ; and thus far the States have respected that sound distinction—have taxed the railway as real estate in the localities where it exists, and the comparatively inconsiderable amount of personal property at the principal place of business of the Company—and have abstained from any attempt to tax interests in fixed property situate in other States, subject to their jurisdiction, and taxable under their laws. The State Governments would be wise to continue the exercise of that comity towards the citizens of each other, so intermingled in their business relations; for the attempt to draw within the taxing power of one State interests in property purely local in another State, would be likely to provoke retaliation in similar cases, and perhaps upon those identical interests of non-residents, and by logical necessity upon all property, business or income of non-residents—which would be discriminating and double taxation on the citizens of the State that should provoke such a conflict of jurisdiction.

The results in the administration of the Pittsburgh, Fort Wayne and Chicago Railway of a liberal policy in developing its capacities to serve the public, in fostering local and through business, and in cheapening transportation, are shown by the following exhibit of its freight traffic in each of the last ten years, the quantity of that traffic, and the rate of charges at which it has been done:

Fort Wayne, Main Line, 468 Miles.

Year ending	Tons Moved one mile.	Rev. per tons per mile. Cts.
December 31, 1859,	58,421,205	1.6
" " 1860,	78,423,319	1.7
" " .861,	111,398,984	1.7
" " 1862,	125,046,905	1.9
" " 1863,	166,570,631	2.
" " 1864,	174,621,870	2.4
" " 1865,	193,789,901	2.5
" " 1866,	233,274,794	2.
" " 1867,	228,791,443	1.9
" " 1868,	307,443,978	1.7

It will be observed that during the inflation incident to civil war and the successive emissions of paper money, the charges were advanced much less than the general increase of prices, or the cost of doing the same quantity of business; and that those charges are now reduced to the rate of about the lowest period after the railway revulsion of 1857 in specie.

The ability to accomplish these results is due to—

1. A cautious and prudent financial policy, combined with a policy of liberal improvement, under which nearly $26,000 per mile has been expended, and yet the capital and debt kept down to about $50,000 per mile, exclusive of sinking funds and investments independent of the property.

2. A cultivation of the natural advantages of a geographical situation which affords peculiar opportunities for freight both ways; so that the per centage of empty cars hauled, including all between local stations, has been as follows:

In 1865.	In 1866.	In 1867.	In 1868.
$26\frac{4}{10}$	$22\frac{5}{10}$	$22\frac{4}{10}$	$17\frac{4}{10}$

3. A concentration of a large traffic on the line, whereby the expenses, which are constant, and the expenses which increase in a less proportion than the traffic, are distributed over a larger business.

The extent of such concentration is illustrated by a comparison with some of the greatest and most prosperous Companies of the country.

The mileage of passenger and freight trains for the year ending December 31, 1868, on 468 miles, was 5,123,324. The mileage of what now constitutes the Lake Shore and Michigan Southern, by their reports to the State of Ohio for the year ending June 30, 1868, and of the Buffalo and Erie, by its report to the State of Pennsylvania for the year ending November 30, 1868, on 540 miles, between Buffalo and Chicago, and on 324 miles of branches, was 5,131,924. The tons of freight on the New York Central, moved one mile for the year ending September 30, 1867, on 593¾ miles, was 362,180,600, or 60,600 tons for each mile ; and on the Fort Wayne last year, on 468 miles, was 307,-443,978, or 65,700 tons for each mile.

4. Cheap fuel, and the very favorable grades and alignment of most of the line, and the advantage of 72 miles less distance from Pittsburgh than from Buffalo to Chicago, with no greater distance from New York to Pittsburgh by Philadelphia, and twelve miles less by Allentown, than from New York to Buffalo.

These have been the principal causes ; but we do not believe that inventions or improvements to cheapen the interchanges between the producer and consumer, which increase the productiveness of human labor, are exhausted. Skillful administration, which is learned by experience, slower trains for freight, with regulated motion, and a larger proportion of load to the dead weight, better materials for rails, and more durable ties, are some of the changes which we may expect. With the growth of transportation by rail, from the lake centres to the seaboard, comes a tendency to longer lines under a united management; which, so long as it does not involve competing lines, or extend beyond the natural centres where the competing lines converge, the public have no occasion to fear.

It is for the public and the investors to recognize each other's rights, even where they are of so subtle a nature as to be within no law or contract. The interest of the public is that capital in such useful enterprises be safe enough to encourage its free investment in them, and at moderate rates of profit. The interest of the investor is to remember that these works are not exclusively for private gain, but involve some element of public character and public trust; and that the benefits of all inventions and improvements, and of successful enterprise and skill applied to them, are to be cheerfully divided with the public.

The present management, undertaken originally mainly by the representatives of the creditors, and conducted under the able and experienced President of the Company, and his principal assistants, for nine and a half years, in closing their participation in the active administration of the property, will practically complete a trust which has been laborious, and from which they are happy to retire, with secured prosperity to the bondholders and stockholders; and they have deemed it proper to submit this exposition of the motives which induced the measures they recommend as beneficial to all parties in interest.

SAMUEL J. TILDEN,

Trustee.

LEASE

OF

THE PITTSBURGH, FORT WAYNE AND CHICAGO

RAILWAY

TO THE

PENNSYLVANIA RAILROAD COMPANY.

THIS INDENTURE, made the seventh day of June, in the year one thousand eight hundred and sixty-nine, between the PITTSBURGH, FORT WAYNE AND CHICAGO RAILWAY COMPANY, duly formed and organized under the laws of the States of Pennsylvania, Ohio, Indiana and Illinois, party of the first part, and the PENNSYLVANIA RAILROAD COMPANY, duly formed and organized under the laws of the State of Pennsylvania, party of the second part:

Whereas, The said party of the first part owns and operates a certain railroad or railway, commonly known as the PITTSBURGH, FORT WAYNE AND CHICAGO RAILWAY, which extends from the City of Pittsburgh, in the State of Pennsylvania, by the way of Crestline, in the State of Ohio, and Fort Wayne in the State of Indiana, to the City of Chicago, in the State of Illinois, and also a certain other railroad or railway which extends from Hudson to Millersburgh in the State of Ohio, and which intersects and connects with the said Pittsburgh, Fort Wayne and Chicago Railway at or near Orville, in the said last mentioned State:

And whereas, The said party of the second part owns and operates a certain railroad, commonly known as the Pennsylvania Railroad, which connects with the said Pittsburgh, Fort Wayne and Chicago Railway in the City of Pittsburgh aforesaid, and extends, by the way of Harrisburgh, to the City of Philadelphia, in the said State of Pennsylvania:

And whereas, The said Pennsylvania Railroad, and the said Pittsburgh, Fort Wayne and Chicago Railway, united, constitute a continuous line of railroad or railway, extending from the City of Philadelphia aforesaid to the City of Chicago aforesaid ; and the parties hereto deem it to be for their common interest, and to the advantage and benefit of each of them, that the said Pittsburgh, Fort Wayne and Chicago Railway should be leased and operated by the said Pennsylvania Railroad Company, for the annual rental hereinafter reserved, and upon and subject to all and singular the terms, agreements, and conditions hereinafter mentioned and set forth :

Now, THEREFORE, THIS INDENTURE WITNESSFTH, That the said party of the first part, for and in consideration of the rents, covenants and agreements hereinafter mentioned, reserved and contained, on the part and behalf of the said party of the second part, to be paid, kept and performed, hath granted, demised and leased, and by these presents doth grant, demise and lease unto the said party of the second part, ALL AND SINGULAR, the railroad or railway now owned and operated by the said party of the first part, and commonly known as the PITTSBURGH, FORT WAYNE AND CHICAGO RAILWAY, which extends from the point of connection with the said Pennsylvania Railroad, in the city of Pittsburgh, in the County of Allegheny and State of Pennsylvania, by the way of Crestline, in the State of Ohio, and Fort Wayne, in the State of Indiana, into the City of Chicago, in the County of Cook and State of Illinois, being a distance of four hundred and sixty-eight (468) miles, more or less ; *and also,* all and singular, the certain other railroad or railway sometimes called the Cleveland, Zanesville and Cincinnati Railway, and sometimes designated as the Akron Branch of the said Pittsburgh, Fort Wayne and Chicago Railway, which is situate in the State of Ohio, and extends from Hudson, in Summit County, to Millersburgh, in Holmes County, in said last mentioned State, being a distance of sixty-one (61) miles, more or less, and runs from thence to the coal fields lying south thereof, in the same county, being a further distance of three and one half miles, more or less, and which said Akron Branch, so called, intersects the said Pittsburgh, Fort Wayne and Chicago Railway, and connects therewith, at or near Orville, in Wayne County, in the said State of Ohio; including in the premises hereby demised, all the railways, ways, and rights of way, and all the depot grounds, and other lands, and all the tracks, bridges, viaducts, culverts, fences, and other structures, and all the depots, station houses, engine houses, car houses, freight houses, wood houses, and other buildings, and all the machine shops, and other shops, appertaining to said Pittsburgh, Fort Wayne and Chicago Railway, and the said Akron Branch, or to either thereof, and now owned by the said party of the first part, and including, also, all locomotives, tenders, passenger, baggage, freight, and other cars, and all other rolling stock and equipment belonging to the said party of the first part, and all rights, privileges and franchises connected with or relating to the said demised railways, or either thereof, or to the construction, maintenance, use or operation of the same; *provided always*, however, that nothing herein contained shall operate to grant or demise, or be construed to include the franchises to be a corporation heretofore granted to the said party of the first part, or the corporators thereof, by the States of Pennsylvania, Ohio, Indiana and Illinois, respectively, or any other right, privilege or franchise which is or may be necessary to preserve the corporate existence or organization of the party of the first part, under its several charters; and all the said franchises to be a corporation, and also all the rights, privileges and franchises last aforesaid are hereby expressly reserved and excepted from these presents:

To have and to hold the said railways and premises, with the appurtenances, unto the said party of the second part, its successors and assigns, from the first day of July, in the year one thousand eight hundred and sixty-nine, for and during and until the full end and term of nine hundred and ninety-nine (999) years then next ensuing, and fully to be complete and ended; the said party of the second part, its successors and assigns, yielding and paying therefor unto the said party of the first part, its successors or assigns, yearly, and every year during the said term hereby granted, the yearly rent hereinafter specified, and keeping and performing all and singular the covenants and agreements hereinafter set forth, to be by the said party of the second part kept and performed; it being understood and provided, however, that in respect to a portion of the said Pittsburgh, Fort Wayne and

Chicago Railway, situate within the City of Chicago, to wit, that portion thereof commencing at the passenger station on Canal street, in said city, and running thence to the junction of said demised railway with the Joilet and Chicago Railroad, at a point at or near the south end of the bridge across the south branch of the Chicago River, in said city, and also in respect to a portion of the said passenger depot on Canal street, these presents are subject to such joint use, right, title and interest as the said party of the first part may have heretofore granted to the Joilet and Chicago Railroad Company, by a deed executed to that company, which is duly recorded in the office of the Register of Cook County, Illinois, and which reference is, for greater certainty, hereby made ; and also that the railroad aforesaid, designated as the Akron Branch, is subject to the lien of a certain deed of trust or mortgage, bearing date on the first day of July, 1865, duly executed and delivered by George W. Cass and wife, and John J. Marvin and wife, to Frederick A. Lane, trustee, to secure the payment of certain bonds issued by the said Cass and Marvin to the said trustee, on account of the purchase money of the said railroad, to the aggregate amount of two hundred thousand dollars, one hundred and forty thousand dollars of which bonds remain outstanding ; and the said party of the second part hereby assumes the payment of, and promises and agrees to pay the principal and interest of said bonds, as the same shall become due, without deduction from the rent herein reserved, and to apply the net earnings of the said Akron Branch, in conformity with the agreement heretofore made by the said party of the first part, to the redemption of the said bonds, until the said bonds shall be fully paid.

ARTICLE FIRST.—The annual rent hereby reserved shall be and consist of the sums in this article specified, and the same shall be paid in lawful money of the United States of America, by the party of the second part to the party of the first part, at the times and places and in the manner following :

First.—One million three hundred and eighty thousand dollars shall be paid in each and every year during the term aforesaid, in quarter-yearly instalments, that is to say : three hundred and forty-five thousand dollars on the first day of April, for and in respect to the quarter ending on the 31st day of

March preceding; three hundred and forty-five thousand dollars
on the first day of July, for and in respect to the quarter ending
on the thirtieth day of June preceding; three hundred and
forty-five thousand dollars on the first day of October, for and in
respect to the quarter ending on the thirtieth day of September
preceding; and three hundred and forty-five thousand dollars
on the first day of January in the next following year, for and
in respect to the quarter ending on the 31st day of December
preceding. The said several instalments shall be paid at the
office or agency, for the time being, of the party of the first part,
in the City of New York, except in any case in which the party
of the first part shall have designated, in writing, a different
place within either of the States in which some part of the said
railway is situate, for the payment of any instalment; in which
case the payment of such instalment shall be made at the place
so designated. Such sum of one million and three hundred and
eighty thousand dollars shall be exclusive of all taxes which are
now or may be at any time hereafter imposed by the Government of the United States, or by or under the authority of the
Government or laws of any of the States of Pennsylvania, Ohio,
Indiana or Illinois, upon the stockholders of the party of the
first part, or any of them, in respect to any capital stock in, or
any dividends upon, or any income derived from capital stock in
the said Pittsburgh, Fort Wayne and Chicago Railway Company, or upon such capital stock, or any income derivable therefrom, so far as such taxes shall be payable or collectable through
the party of the first part, or the party of the second part, or any
officer or agent thereof, or shall be in any lawful manner required
to be collected or paid through either of the said parties, or any
officer or agent thereof, before the actual receipt of such dividends by such stockholders; it being the true intent and meaning of these presents that the said sum of one million three hundred and eighty-three thousand dollars shall at all times hereafter be and remain applicable by the party of the first part as a
dividend fund for the stockholders of the said party of the first
part, without any deduction or abatement on account of any
such tax, and that every such tax shall be paid by the party of
the first part in addition thereto.

Secondly.—In addition to the aforesaid sum of one million
three hundred and eighty thousand dollars a further sum shall

be paid by the party of the second part to the party of the first part, in each and every year, which shall be sufficient to provide for and pay all instalments of interest and all instalments of sinking fund, which may become payable during such year, by the party of the first part, upon bonds issued or assumed by the said party, to wit :

1. A sum equal to the annual interest on the bonds of the party of the first part, secured by a first deed of trust or mortgage, to John Ferguson and Samuel J. Tilden, bearing date on the first day of March, 1862, and amounting in the aggregate to five millions and two hundred and fifty thousand dollars ; and on bonds of the party of the first part, secured by a second deed of trust or mortgage, to John Ferguson and Samuel J. Tilden, bearing the same date, and amounting in the aggregate to five millions one hundred and sixty thousand dollars, and to the instalments payable during each rear to the sinking funds for the redemption of the said two series of bonds ; that is to say, the sum of eight hundred and thirty-two thousand and eight hundred dollars shall be paid in each and every year during the term aforesaid, until the said bonds shall be fully discharged and paid by the operation of the said sinking funds. Such sum shall be paid in monthly instalments of one equal twelfth part thereof, being the amount of sixty-nine thousand and four hundred dollars, to the party of the first part, at the office or agency of the party of the first part within the City of New York, designated for the time being the payment of interest accruing upon the said bonds ; and each such instalment of sixty-nine thousand and four hundred dollars shall be paid on or before the twenty-fifth day of each month in every such year.

2. A further sum equal to the annual interest on the bonds of the party of the first part, designated as Income Bonds, secured by a third deed of trust or mortgage to John Ferguson and Samuel J. Tilden, bearing the same date, and amounting in the aggregate to two million of dollars ; that is to say, the sum of one hundred and forty thousand dollars shall be paid in each and every year of the aforesaid term, unless and until the principal of the said bonds shall have been fully paid by the party of the second part. Such payment shall be made to the party of the first part at the office or agency last aforesaid, in two semi-annual instalments of seventy-two thousand dollars each, payable

on or before the twenty-eighth day of March and twenty-eighth day of September in each and every year.

3. A further sum, equal to the interest and sinking fund on certain bonds of the Ohio and Pennsylvania Railroad Company, commonly known as the Bridge Bonds thereof, and amounting in the aggregate to about one hundred and fifty-three thousand and one hundred dollars, which are payable on the first day of May, 1876, and bear interest at the rate of seven per centum per annum, payable on the first days of May and November in each year, and the payment of which has been assumed by the party of the first part, and is secured by a certain deed of trust or mortgage, bearing date on the sixth day of May, 1856, made by the said Ohio and Pennsylvania Railroad Company to Thomas T. Firth and Reuben Miller, Jr., and whatever additional sum the party of the first part may be bound to pay thereon, as a tax, to the State of Pennsylvania, accruing after the first day of July, 1869, which said sum, equal to the interest and sinking fund on said Bridge Bonds, shall be paid to the said party of the first part, by the payment to the trustees, for the time being, under said last mentioned deed of trust or mortgage, of the sum of two thousand dollars per month, to be paid to such trustees on the first day of each and every month until the said Bridge Bonds shall be fully paid and satisfied, as provided in the said deed of trust or mortgage.

4. A further sum, equal to the interest on certain bonds of the Pittsburgh, Fort Wayne and Chicago Railroad Company, known as Construction Bonds, which bear date on the first day of January, 1857, amounting in the aggregate to one hundred thousand dollars, are payable in thirty years from date, and bear interest at the rate of seven per centum per annum, payable semi-annually on the first days of January and July of each year, which said bonds were delivered in payment for certain depot grounds in the City of Pittsburgh, and the payment whereof has been assumed by the said party of the first part, and is secured by a certain deed of trust or mortgage on said depot grounds, made to Samuel Haight and wife; that is to say, the sum of seven thousand dollars shall be paid by the party of the second part, in each year, to the party of the first part, at its office or agency in the City of New York, in instalments of three thousand and five hundred dollars each, payable semi-annually on the twenty-

eighth day of June and the twenty-eighth day of December in each year, for the purpose of paying to the holders, for the time being, of the said last mentioned bonds the interest thereon, as they shall semi-annually become due, until the principal of the said bonds shall be fully paid by the said party of the second part.

5. A further sum, equal to the interest on certain bonds of the party of the first part, designated as Equipment Bonds, which bear date the first day of March, 1869, and amount in the aggregate to one million of dollars; that is to say, eighty thousand dollars, and the income tax of the Government of the United States thereon shall be paid in each and every year of the said term until the principal of the said bonds shall be fully paid by the party of the second part; which sum of eighty thousand dollars and the aforesaid tax thereon, shall be paid to the party of the second part, at the office or agency last aforesaid, in semi-annual instalments of forty thousand dollars each, and the aforesaid tax thereon; on or before the twenty-eighth day of February and August in each and every year.

Provided, however, and it is hereby agreed, in respect to the several instalments of interest and sinking fund by this article made payable at the office or agency of the party of the first part in the City of New York, that if the party of the first part shall give to the party of the second part reasonable notice, in writing, for the payment of any such instalment at a place within either of the States within which the said Pittsburgh, Fort Wayne and Chicago Railway, or any part thereof, is situate, such instalment shall be paid at that place.

And the said party of the first part hereby promises and agrees, that if the several sums of money herein provided to be paid, on account of interest and sinking fund on said bonds, shall be paid to the said party of the first part as herein agreed, the same shall be applied to the payment of said interest and sinking fund, and due evidence of such payments and application be furnished to the said party of the second part when required; and that if the said party of the first part, receiving the several sums of money aforesaid, shall in any year fail to apply the same in payment of such interest and sinking fund, the interest or sinking fund so remaining unpaid may be paid to the said party of the second part directly to the parties to whom the same may be due from the said party of the first part, and the amount thereof charged to the said party of the first part on account of the rent herein reserved.

ARTICLE SECOND.—The party of the first part hereby agrees, in respect to the aforesaid issue of Income or Third Mortgage Bonds, which are payable at the pleasure of the party of the first part, after the first day of July, 1912, that the party of the first part shall and will exercise such option as it lawfully may by virtue of the agreements expressed in the said bonds and the deed of trust whereby they are secured, at such time or times and in such manner as the party of the first part may be in writing requested by the party of the second part ; *provided, however*, that in case the party of the second part shall request such option to be exercised in any manner which shall involve the calling in for payment, or the paying of any of the said Income Bonds, the said party of the second part shall, at the time of making such request, furnish to the party of the first part the money for such payment ; and the party of the first part further agrees, that it will not, at any time during the term aforesaid, exercise the aforesaid option to call in the said Income Bonds for payment, except upon the request in writing of the party of the second part.

And the party of the first part hereby promises and agrees to and with the party of the second part, that the said party of the first part shall and will, upon the request in writing of the said party of the second part, consent and agree to the extension, for such period as shall be specified in said request, of the time of payment of the principal of said Equipment Bonds, or shall, on the like request, issue, in substitution for said Equipment Bonds, other bonds to an amount not exceeding the aggregate amount of the said Equipment Bonds then outstanding, which other bonds shall bear not exceeding the same rate of interest as the said Equipment Bonds now bear, and shall be payable at such time, after the maturity of the said Equipment Bonds, as shall be specified in such request for the issue thereof; and that the said party of the first part shall and will execute, at the cost and expense of the party of the second part, such instruments, and do such other acts as may be necessary and proper for the purpose of such extension or substitution of bonds ; *provided*, that the party of the second part shall continue to pay, and the said party of the second part hereby agrees to continue to pay the said party of the first part, as a portion of the rent reserved under this lease, the annual interest on any and all bonds which shall be issued in substitution for said Equipment Bonds as aforesaid, in the same manner as hereinbefore agreed in respect to said Equipment Bonds.

And it is further agreed, that the party of the second part shall bear the expenses and be entitled to the savings, if any, resulting from any such extension or substitution of said bonds ; and the said party of the second part hereby agrees to provide for the payment of, and to pay the said Equipment Bonds at the maturity thereof, if the same shall not be extended or substituted as aforesaid, or in case of an extension of the time of payment of said bonds, or the issue of other bonds in substitution therefor, to provide for the payment of, and to pay the bonds so extended or substituted at the maturity thereof, and also to provide for and pay the principal of the aforesaid Construction Bonds of the said Pittsburgh, Fort Wayne and Chicago Railroad Company, to the amount of one hundred thousand dollars, which have been assumed by the party of the first part. And the party of the second part further agrees, that it shall and will, at any and all times hereafter, indemnify and save harmless the party of the first part, of, from and against any payment, cost or liability whatsoever, in respect to the same, or in respect to the said Equipment Bonds, as well as in respect to the principal of the said Income or Third Mortgage Bonds.

ARTICLE THIRD.—It being provided herein, that the term hereby granted shall begin on the first day of July, in the year one thousand eight hundred and sixty-nine, it is hereby agreed and declared that the rental provided to be paid by Article First of these presents shall commence to run on that day, and shall be payable for the first six months, beginning with the said first day of July, and ending with the thirty-first day of next December thereafter, in the same instalments, at the same times and places, and in the same manner as such rental would be payable for the corresponding last six months of the next succeeding year under the provisions of this instrument, except as to the interest payment on the twenty-eight days of August and September, that is to say : the first quarterly instalment of the annual sum required by the first subdivision of Article First of these presents, being three hundred and forty-five thousand dollars, shall be paid on the first day of October, 1869, in conformity to the provisions of the said subdivision, and the second quarterly instalment of the said annual sum, being of the same amount, shall be paid on the first day of January, 1870, in like conformity to the said provisions; and the several payments, specified in the second subdivision of Article First of these presents, to be made for the purpose of providing for the instalments of interest and sinking funds on the bonds mentioned therein, shall be paid as follows:

1. Sixty-nine thousand and four hundred dollars on the twenty-eighth day of each of the months of July, August, September, October, November and December.

2. Thirty-five thousand dollars on the twenty-eighth day of September.

3. Two thousand dollars in each month, according to the requirements of the deed of trust or mortgage to Thomas T. Firth and Reuben Miller, Jr., hereinbefore mentioned.

4. Three thousand and five hundred dollars on the twenty-eighth day of December.

5. Thirteen thousand three hundred and thirty-three $\frac{33}{100}$ dollars, and the Government tax thereon, on the twenty-eighth day of August.

ARTICLE FOURTH.—The party of the second part, for itself, its successors and assigns, hereby covenants, promises and agrees to and with the party of the first part, its successors and assigns, in consideration of the execution to it of this lease, that the said party of the second part, its successors and assigns, shall and will yearly, and in each and every year of and during the term aforesaid, well and truly pay, or cause to be paid, unto the said party of the first part, its successors or assigns, the yearly rent hereinbefore reserved, to wit, the several sums provided by Article First of these presents, to be paid by the said party of the second part to or on account of the said party of the first part, in the instalments, at the times, and in the manner in the said Article mentioned.

ARTICLE FIFTH.—The party of the second part shall and will, at its own proper cost and expense, and without deduction from the rent aforesaid, operate and run the said demised railways at all times during the said term, in the same manner as the said party of the first part, as the owner thereof or otherwise, is now or shall or may at any time hereafter be required by law to do; and the said party of the second part shall and will, at its own proper cost and expense, and without deduction from the rent aforesaid, at all times during the said term, maintain, preserve and keep the railways and premises hereby demised, and every part of the same, in thorough repair, working order and condition, and supplied with rolling stock and equipment, so that the business of said demised railways shall be preserved, encouraged, and developed, and that the same shall at all times be done with safety and expedition, and the public be accommodated in respect thereto with all practicable conveniences and facilities, and that all future growth of such business, as the same may arise or be reasonably anticipated, shall be fully provided for and secured. And the party of the second part hereby promises and agrees to and with the party of the first part that the said party of the second part shall and will, at its own proper cost and expense, and without deduction from the rent aforesaid, from time to time, and whenever needed, during the term aforesaid, do or cause to be done to and upon the said demised railways and premises any and all repairs, replacements and renewals, and also, in the manner expressed in the sixteenth article of these presents, any and all additions, constructions and improvements which may be reasonably required for the purposes aforesaid, and provide thereon such rolling stock and equipment, and other facilities, as shall or may be reasonably required for the purposes aforesaid; and that the said party of the second part shall and will use all reasonable efforts to maintain, develop and increase all the business of the said railways.

ARTICLE SIXTH.—The said party of the second part shall and will, from time to time, and as often as the same shall become due, also pay and discharge, without deduction from the rent aforesaid, any and all taxes, assessments, duties, imports and charges whatsoever, which shall or may be levied, assessed or imposed during the said term, by any government or lawful authority whatever, upon the said demised railways and premises, or upon any part thereof, or upon any business or earnings or income of the same, or upon the party of the first part, with respect to the said demised railways and premises, or any part thereof, or any business or earnings or income of the same, or upon the party of the first part, for or with respect to any money which shall be paid, or which shall become payable to the said party of the first part, as or on account of the rent hereinbefore reserved in Article First of these presents, or any money which shall be paid or become payable to the said party of the first part, in pursuance of these presents, or with respect to any interests or rights under these presents, or upon such money, interests or rights; or which shall or may be levied, assessed or imposed upon any stockholder or stockholders of the party of the first part in respect to capital stock in the said party of the first part, or in respect to any dividends upon or any income from such capital stock, or upon such dividends, or income, or capital stock, by or under any governmental authority which is or may be exercised over any territorial jurisdiction within which any part of the said demised railways and premises is or may be situate, except such as shall be collected by or under the laws of any State from its own citizens personally, without any action upon or through, or any intervention or service of either of the parties hereto, or any officer or agent thereof, it being the true intent and meaning of these presents that all governmental charges upon the aforesaid property, or the stockholders, with respect to such property or income therefrom, which may be imposed by or under any governmental authority capable of enforcing such charges against or through the said property, or the corporation owning or leasing the same, shall be assumed and satisfied by the party of the second part, however the forms thereof may change during the term of these presents; but that the said party of the second part shall not be or become liable to pay any tax imposed by any law of any of the States within which the said Pittsburgh, Fort Wayne and Chicago Railway, or any part thereof, is situate, or by any law of any other State, upon citizens of such State personally, in respect to stock held by them, or dividends or income derived by them therefrom, which shall be collected from such citizens personally, without any action upon or through, or any intervention or service of the party of the first part, or any officer or agent thereof; nor shall anything herein contained be construed to render the party of the second part liable to pay any tax imposed by the Government of the United States, or of any State, especially upon income derived from interest on the bonds of the party of the first part, except as to certain Equipment Bonds and certain Bridge Bonds hereinbefore mentioned.

ARTICLE SEVENTH.—*Whereas*, on or about the twenty-ninth day of June, A. D. 1865, the New Castle and Beaver Valley Railroad Company executed to the party of the first part a certain indenture of agreement or lease, bearing date on said last mentioned day, in and by which the New Castle and Beaver Valley Railroad Company let, leased and demised the New Castle and Beaver Valley Railroad, therein described, to the said party of the first part, for the term of ninety-nine years, commencing on the first day of July, A. D. 1865, upon the terms and conditions, and in consideration of the rents in said indenture of agreement or lease mentioned and reserved.

Now, therefore, this Indenture witnesseth, that for and in consideration of the premises, and of the sum of one dollar to the said party of the first part duly paid, and of the covenants and agreements on the part and behalf of the said party of the second part herein contained, the said party of the first part has sold, and does hereby sell, assign, transfer and set over unto the said party of the second part the said agreement or indenture of lease, bearing date on the twenty-ninth day of June, 1865, made by the said New Castle and Beaver Valley Railroad Company, and all the right, title and interest of the said party of the first part therein, or in or to the railroad and premises thereby demised, under or by virtue thereof; *provided, always,* that the said party of the second part hereto shall well and truly pay, or cause to be paid, the rental reserved by said lease, and well and truly keep and perform the several matters and things therein provided to be kept or performed by the said party of the first part hereto, and shall at all times indemnify and save harmless the said party of the first part hereto, of, from and against any and all claims and demands whatsoever, arising under or by virtue of said agreement or lease, or out of or in connection with the possession, management or operation of the railroad and premises therein mentioned, or the business of the same. And, the said party of the second part, in consideration of the premises, has promised and agreed, and does hereby promise and agree to and with the said party of the first part, that the said party of the second part shall and will well and truly pay the moneys in or by the said agreement or lease agreed to be paid, by the said party of the first part hereto, to the said New Castle and Beaver Valley Railroad Company, as rent or otherwise, and that the said party of the second part will do and perform every act and thing which the said party of the first part hereto has agreed to do or perform, in or by said agreement or lease; and the said party of the second part hereby assumes and takes the place of the said party of the first part under said agreement or lease, and promises and agrees to save and keep harmless and indemnified the said party of the first part, of, from and against any and all claims, demands and liabilities whatsoever, which can or may arise against the said party of the first part, from, under, or on account of the said agreement or lease, or the railroad or premises therein mentioned, or the possession, operation, management or business of the same.

ARTICLE EIGHTH.— *Whereas*, on or about the twenty-second day of May, A. D. 1869, the Lawrence Railroad Company executed to the party of the first part a certain indenture of agreement or lease, bearing date on said last mentioned day, in and by which indenture of agreement or lease the said Lawrence Railroad Company let, leased, and demised its railroad to the party of the first part for the term of ninety-nine years, commencing on the first day of July, A. D. 1869, upon the terms and conditions, and in consideration of the rents in said agreement or lease mentioned and reserved:

Now, therefore, this Indenture witnesseth, that for and in consideration of the premises, and of the sum of one dollar to the said party of the first part duly paid, and of the covenants and agreements on the part and behalf of the said party of the second part herein contained, the said party of the first part has sold, and does hereby sell, assign, transfer, and set over unto the said party of the second part, the said agreement and indenture of lease bearing date on the twenty-second day of May, 1869, made by the said Lawrence Railroad Company, and all the right, title, and interest of the said party of the first part therein, or in the railroad and premises thereby demised, under or by virtue thereof; *provided, always*, that the said party of the second part hereto shall well and truly pay the rental reserved by said lease, and well and truly keep and perform the several matters and things therein provided to be kept or performed by the said party of the first part hereto, and shall at all times indemnify and save harmless the said party of the first part hereto, of, from and against any and all claims and demands whatsoever arising under or by virtue of said agreement or lease, or out of or in connection with the possession, management, or operation of the railroad and premises therein mentioned, or the business of the same. And the said party of the second part, in consideration of the premises, has promised and agreed, and does hereby promise and agree, to and with the said party of the first part, that the said party of the second part shall and will well and truly pay the moneys in or by the said agreement or lease agreed to be paid by the said party of the first part hereto, to the said Lawrence Railroad Company, as rent or otherwise, and that the said party of the second part will do and perform every act and thing which the said party of the first part hereto has agreed to do or perform in or by said agreement or lease; and the said party of the second part hereby assumes and takes the place of the said party of the first part under said agreement or lease, and promises and agrees to save and keep harmless and indemnified the said party of the first part, of, from and against any and all claims, demands and liabilities whatsoever which can or may arise against the said party of the first part from or under or on account of the said agreement or lease, or the railroad or premises therein mentioned, or the possession, operation, management or business of the same.

ARTICLE NINTH.— *Whereas,* on or about the twenty-second day of May, A. D. 1869, the Massillon and Cleveland Railroad Company executed to the party of the first part a certain indenture of agreement or lease, bearing date on said last mentioned day, in and by which the said Massillon and Cleveland Railroad Company let, leased and demised the Massillon and Cleveland Railroad, therein described, to the said party of the first part, for the term of ninety-nine years, commencing on the first day of July, 1869, upon the terms and conditions, and in consideration of the rents in said agreement or lease mentioned and reserved:

Now, therefore, this Indenture witnesseth, that for and in consideration of the premises, and of the sum of one dollar to the said party of the first part duly paid, and of the covenants and agreements on the part and behalf of the said party of the second part herein contained, the said party of the first part has sold, and does hereby sell, assign, transfer and set over unto the said party of the second part the said agreement or indenture of lease, bearing date on the twenty-second day of May, 1869, made by the said Massillon and Cleveland Railroad Company, and all the right, title and interest of the said party of the first part therein, or in the railroad and premises thereby demised, under or by virtue thereof ; *provided, always,* that the said party of the second part hereto shall well and truly pay the rental reserved by said lease, and well and truly keep and perform the several matters and things therein provided to be kept or performed by the said party of the first part hereto, and shall at all times indemnify and save harmless the said party of the first part hereto, of, from and against any and all claims and demands whatsoever arising under or by virtue of said agreement or lease, or out of or in connection with the possession, management or operation of the railroad and premises therein mentioned, or the business of the same. And the said party of the second part, in consideration of the premises, has promised and agreed, and does hereby promise and agree, to and with the said party of the first part, that the said party of the second part shall and will well and truly pay the moneys in or by said agreement or lease agreed to be paid by the said party of the first part hereto, to the said Massillon and Cleveland Railroad Company, as rent or otherwise ; and that the said party of the second part will do and perform every act and thing which the said party of the first part hereto is or can be required to do or perform, or which it has agreed to do or perform, in or by said agreement or lease ; and the said party of the second part hereby assumes and taxes the place of the said party of the first part under said agreement or lease, and promises and agrees to save and keep harmless and indemnified the said party of the first part of, from and against any and all claims, demands and liabilities whatsoever, which can or may arise, against the said party of the first part, from or under, or on account of the said agreement or lease on the railroad or premises therein mentioned, or the possession, operation, management or business of the same.

ARTICLE TENTH.— *Whereas,* the said party of the first part, heretofore, to wit, on or about the fifteenth day of December, in the year 1862, made and entered into a certain contract or agreement, bearing date on said last mentioned day, with the Cleveland and Pittsburgh Railroad Company, in respect to the operation and business of the respective roads of said companies, which contract or agreement was subsequently, to wit, on or about the sixteenth day of February, 1866, duly awarded by a further contract or agreement, bearing date on said last mentioned day, as by reference to said original contract, and to said amendatory contract, will more fully and at large appear; which said contracts with the Cleveland nd Pittsburgh aRailroad Company are still in force and operation:

Now this Indenture witnesseth, that for and in consideration of the premises, and of the sum of one dollar to it duly paid, the said party of the first part has sold, and does hereby sell, assign, transfer and set over unto the said party of the second part, all the right, title, interest, claim and demand whatsoever of the said party of the first part, in, to and under said contracts or agreements with the Cleveland and Pittsburgh Railroad Company: *provided, always,* that the said party of the second part shall at all times hereafter indemnify and save harmless the said party of the first part, of, from and against the said contracts, and each thereof, and all claims and demands thereunder; and the said party of the second part hereby covenants, promises and agrees to and with the said party of the first part, that the said party of the second part shall and will, at all times hereafter, hold, save and keep harmless and indemnified the said party of the first part, of, from and against said contracts with the Cleveland and Pittsburgh Railroad Company, and all balances of account, payments, charges, claims and liabilities whatsoever, which shall or may be, or at any time hereafter arise or become payable under or by virtue of said contracts, or either of them, accruing after the first day of July, 1869; and the said party of the second part hereby promises and agrees to and with the said party of the first part, that the said party of the second shall and will observe and keep the said contracts or agreements with the Cleveland and Pittsburgh Railroad Company, according to the true intent and meaning thereof, and shall and will do and perform every act and thing which the said party of the first part, in or by said contracts, or either of them, is or may be required to do or perform, and which the said party of the second part lawfully may do or perform, instead of the said party of the first part under said contracts; and that as to any matter or thing agreed to be done by the said party of the first part, in or by said contracts, or either of them, which under said contracts cannot be lawfully done by the said party of the second part, the said party of the first part shall be, and it is hereby authorized and empowered to do and perform the same, and the said party of the first part promises and agrees that it shall do and perform the same, provided the said party of the second part bear and pay all costs and expenses incident thereto.

And the said party of the first part promises and agrees, to and with the said party of the second part, that the said party of the first part shall do every act to be by it done for the appointment of the Executive Committee, provided to be appointed by the fourth section of said contract, bearing date on the fifteenth day of December, 1862, and that it will nominate as a member of such Committee such member of the Board of Directors as the said party of the second part may in writing designate, whenever, pursuant to said contract, it shall be agreed to place on said Committee, in addition to the Presidents of the respective companies, a Director of each company ; and that the said party of the first part will at any time, on the request of the party of the second part, agree to the addition of such Directors to said Committee; and that in case of any disagreement arising between the members of said Committee in regard to the true, proper or legal meaning of said contract, or the exercise of powers under the same, the party of the first part shall name as the arbiter or additional member of said Committee, provided to be appointed by the fifth section of said original contract in such contingency, such competent and disinterested person, familiar with railway management, as the said party of the second part shall, in writing, designate for that purpose. And the said party of the second part promises and agrees, to and with the said party of the first part, that the Executive Committee, so to be appointed under said contract, shall and may exercise, in respect to the said demised railway, all and singular the powers and authorities provided to be exercised by such Committee, under or by virtue of said contracts, and that the said party of the second part will pay to the members of said Committee, appointed by the party of the first part, a reasonable compensation for all services to be rendered on said Committee, and indemnify and save harmless the party of the first part from all charges in respect thereto.

And the said party of the first part promises and agrees, to and with the said party of the second part, that the Executive Committee of the said party of the first part shall appoint as the the General Superintendent provided to be appointed by the fifth section of the said contract bearing date on the 15th day of December, 1862, and as the General Freight Agent and as the General Ticket Agent provided to be appointed by the sixth section of said contract, such person as shall from time to time be designated by the said party of the second part to fill such positions. And the said party of the second part promises and agrees, that such Superintendent, General Freight Agent and General Ticket Agent shall each possess, and be authorized to exercise, in respect to the said demised railways, and premises, all the powers and authorities specified in said contracts, to be possessed or exercised by such persons respectively.

ARTICLE ELEVENTH.— *Whereas*, heretofore, to wit, on or about the 17th day of May, in the year 1867, the St. Louis, Alton and Terre Haute Railroad Company, at the instance and request of the party of the first part hereto and others, executed a certain instrument in writing, dated on said last mentioned day, which purported to be made between the Terre Haute and Indianapolis Railroad Company as party of the first part, and the St. Louis, Alton and Terre Haute Railroad Company as party of the second part, and to be a contract for the operation of the main line and Alton Branch of the St. Louis, Alton and Terre Haute Railroad, as by reference to said instrument will more fully and at large appear ;

And whereas, the said operating contract was executed by said St. Louis, Alton and Terre Haute Railroad Company upon the promises and guaranty of the parties requesting the execution of the same as aforesaid, that the said Terre Haute and Indianapolis Railroad Company would, on or before the first day of July next thereafter, execute the same on its part, or, in default thereof, that the said party of the first part hereto, and others, would provide another company, owning or constructing a line of railroad from Indianapolis to Terre Haute, to execute the said operating contract and assume the obligations purporting by the terms thereof to be assumed by the said Terre Haute and Indianapolis Railroad Company, as by reference to said agreement or guaranty will more fully and at large appear ;

And whereas, the said Terre Haute and Indianapolis Railroad Company failed and omitted to execute said operating contract within the time limited for the execution thereof, and thereupon the party of the first part hereto, and others, procured to be organized a new corporation, called the Indianapolis and St. Louis Railroad Company, with authority to construct a railroad from Indianapolis to Terre Haute, and requested the said St. Louis, Alton and Terre Haute Railroad Company to accept the said new corporation as the party of the first part to the said operating contract in the place and stead of the said Terre Haute and Indianapolis Railroad Company, which had failed to execute the same ; *and whereas,* the said St. Louis, Alton and Terre Haute Railroad Company accepted such new company in the place and stead of said Terre Haute and Indianapolis Railroad Company, upon the premises and guaranty of the said party of the first part hereto, and others, that the said Indianapolis and St. Louis Railroad Company would keep, observe and perform the conditions, covenants and provisions of the said operating contract purporting to be made by the Terre Haute and Indianapolis Railroad Company, in the same manner, and with the same effect, as if the said Indianapolis and St. Louis Railroad Company had been the original party of the first part to the said operating contract, as by reference to a certain agreement, bearing date on the 11th day of September, 1867, made by the party of the first part hereto, and others, with the said St. Louis, Alton and Terre Haute Railroad Company, will more fully and at large appear ;

And whereas, the said Indianaoplis and St. Louis Railroad Company, by an instrument, in writing, bearing date on the said eleventh day of September, 1867, assumed, adopted and became liable to carry out the said operating contract, bearing date on the seventeenth day of May, in the year 1867, and purporting to be entered into by the Terre Haute and Indianapolis Railroad Company, as by reference to said instrument, assuming said operating contract, will more fully and at large appear ;

And whereas, under and by virtue of the said agreement of guaranty, bearing date on the eleventh day of September, 1867, the said party of the first part hereto is liable to pay the equal one-third part of any and all damages which may arise to the said St. Louis, Alton and Terre Haute Railroad Company, for or on account of any default of the said Indianapolis and St. Louis

Railroad Company, in the performance of the said operating contract, the performance whereof is guaranteed as aforesaid :

Now, therefore, this Indenture further witnesseth, that the said party of the second part has agreed, and does hereby covenant, promise and agree, to and with the said party of the first part, that the said party of the second part shall and will bear, pay and discharge all liabilities, claims and demands whatsoever, which shall or may arise against the said party of the first part, under or by virtue or on account of the said agreements or contracts with the St, Louis, Alton and Terre Haute Railroad Company, or either of them, and that the said party of the second part hereto will faithfully carry out and perform any and all stipulations and agreements in said contracts set forth to be carried out or performed by the said party of the first part hereto, and that the said party of the second part hereto will at all times keep and save harmless and indemnified the said party of the first part hereto, of, from and against any and all liabilities, claims and demands whatsoever which shall or may arise under or by virtue or on account of the said contracts with the St. Louis, Alton and Terre Haute Railroad Company.

And whereas, on or about the first day of May, in the year 1869, the said party of the first part hereto, and the Cleveland, Columbus, Cincinnati and Indianapolis Railroad Company entered into a certain contract between themselves, bearing date on that day, relating to the management and operation of the railroad mentioned in said operating contract, as by reference to said contract between the said party of the first part and the said Columbus, Cincinnati and Indianapolis Railroad Company will more fully and at large appear.

Now, therefore, the said party of the second part hereby promises and agrees to carry out the said last-mentioned contract, according to the true intent and meaning thereof, on the part and behalf of the party of the first part, and to save and keep the said party of the first part harmless and indemnified from and against any and all claims, demands and liabilities whatsoever, which shall or may arise under or on account thereof.

And whereas, the said party of the first part, for the purpose of carrying into effect, according to the true intent and meaning thereof, the aforesaid contracts with the St. Louis, Alton and

Terre Haute Railroad Company, and procuring the construction of the said Indianapolis and St. Louis Railroad, as thereby contemplated, has subscribed, or agreed to subscribe, and pay for three hundred thousand dollars in par value, or six thousand dollars of the par value of fifty dollars per share, of the capital stock of the said Indianapolis and St. Louis Railway Company, to be paid for to the said Company in such instalments as shall be called for by the Board of Directors thereof, and has also agreed to take and pay for, or procure to be negotiated and paid for, at ninety per cent. of their par value, five hundred thousand dollars in amount of bonds, to be issued by the said Indianapolis and St. Louis Railroad Company, and secured by mortgage on the railroad to be constructed by said company, which said bonds so subscribed for, or agreed to be subscribed for, are to be paid for by the said party of the first part at the price aforesaid, at such time and in such manner as the Board of Directors of the said Indianapolis and St. Louis Railroad Company shall require.

Now, therefore, this Indenture further witnesseth, that the said party of the second part has agreed, and does hereby promise and agree, to and with the party of the first part, that the said party of the second part shall and will, upon demand by the said party of the first part, purchase and take from the said party of the first part six thousand shares of the stock of the said Indianapolis and St. Louis Railroad Company of the par value of fifty dollars per share, and pay to the said party of the first part for the same the cost thereof to the said party of the first part, according to the terms of the subscription for, or agreement to take the same, and also that the said party of the second part shall and will, upon demand by the said party of the first part, purchase and take from the said party of the first part from time to time, and as the same shall be received by the party of the first part, pursuant to the aforesaid subscription or agreement to subscribe, the aforesaid bonds of the Indianapolis and St. Louis Railroad Company, to the amount of five hundred thousand dollars, and pay to the said party of the first part, for the same, the cost thereof to the said party of the first part, at the aforesaid rate of ninety per cent. upon the par value thereof; and that the said party of the second part will indemnify and save harmless the party of the first part from all loss and damage, by reason of such subscriptions, or agreements to subscribe, for stock and bonds, or either of them.

ARTICLE TWELFTH.— *Whereas*, the said party of the first part has also heretofore made and entered into other contracts and agreements, in relation to the business of said demised railways, and it is the intent and meaning of these presents that all contracts relating thereto should be assumed and carried out by the party of the second part, that the benefit thereof should enure and the liability thereof rest on the said party of the second part, in the place and stead of the party of the first part :

Now, therefore, the said party of the first part hereby sells, assigns, transfers and sets over unto the said party of the second part, all the right, title and interest of the said party of the first part, in and under all contracts and agreements, in relation to the business to be done on the said demised railways, or any part thereof not hereinbefore specifically transferred, and the said party of the second part hereby assumes the performance of the same, on the part and behalf of the said party of the first part, according to the true intent and meaning thereof ; and the said party of the second part hereby promises, covenants and agrees to and with the said party of the first part, that the said party of the second part shall and will, at all times hereafter, save and keep harmless and indemnified the said party of the first part, its successors and assigns, of, from and against all costs, damages, expenses, liabilities, claims and demands whatsoever, which may exist or shall or may arise under the said contracts and agreements, or either thereof.

ARTICLE THIRTEENTH.— *Whereas*, a negotiation is now pending between the party of the first part and the Grand Rapids and Indiana Railroad Company, for the purpose of securing to the party of the first part or its assigns, any and all such business as may be contributory by the said Grand Rapids and Indiana Railroad Company, to the said party of the first part, or its assigns, in the connection of the respective roads of said companies, and thereby the better to develop and increase the business of each Company :

Now, therefore, this Indenture further witnesseth, that the said party of the first part has agreed, and does hereby promise and agree to and with the said party of the second part, that the said party of the first part shall and will assign and transfer to the said party of the second part, such contract, if any, as shall or may be concluded by the said party of the first part with the said Grand Rapids and Indiana Railroad Company, for the purposes aforesaid, in consideration that the said party of the second part shall assume and undertake the performance of such contract, and indemnify and save harmless the said party of the first part, of, from and against the same, and any and all claims, demands and liabilities whatsoever which can or may arise thereunder, and the said party of the second part hereby agrees to assume such contract, and to indemnify and save harmless the said party of the first part, of from and against the same, and all claims and demands thereunder, provided such contract or proposed contract shall not involve the guaranty of an amount exceeding four millions of dollars of first mortgage bonds of the said Grand Rapids and Indiana Railroad Company, the whole issue of which first mortgage bonds shall not exceed eight millions of dollars, and which said issue shall be further secured by a deed of trust of one million of acres of land, and that a majority in amount of the capital stock of the said Grand Rapids and Indiana Railroad Company shall be deposited in trust until the principal and interest of the said guaranteed bonds shall be provided for, and the guaranty satisfied and canceled, and that the form of such contract or proposed contract shall be submitted to and approved by the Board of Directors of the party of the second part.

ARTICLE FOURTEENTH.—The party of the second part shall and will, at all times during the term aforesaid, allow and pay to the party of the first part, without deduction from the rent aforesaid, and as part of the necessary expenses of carrying on the business of said demised railways, the sum of seven thousand dollars per annum, to enable the said party of the first part to pay to an agency in the City of New York, or elsewhere, proper compensation for the services to be rendered by such agency in the payment of coupons, representing interest on the bonds of the party of the first part hereinbefore mentioned, as the same from time to time mature, and of the principal of said bonds when the same shall become due, and in the transfer of the stock of the said party of the first part, and the issue of certificates therefor, and in the payment of dividends to the stockholders; and the said party of the second part shall also pay to the said party of the first part the further sum of two thousand dollars per annum, to enable the said party of the first part to pay the expenses of a registry of transfers of the stock of the said party of the first part, and of the certificates issued therefor; which said sums of seven thousand and two thousand dollars shall be paid to the said party in four equal quarter-yearly instalments of two thousand two hundred and fifty dollars each, payable on the first days of January, April, July and October, in each and every year of the said term; provided, nevertheless, that it is the intention of the parties that these expenses shall exist only to the extent and for the period during which services of such value shall be required, and that for the purpose of adapting the same to events as they change, the aforesaid provisions of this Article shall be subject to modification from time to time, by the agreement of the parties to these presents.

And for the purpose of enabling the said party of the first part to maintain and preserve its corporate organization, and pay the salaries of its officers, the said party of the second part shall allow and pay to the said party of the first part, during the said term, the further sum of ten thousand dollars per annum, to be paid in four equal quarter-yearly instalments of two thousand five hundred dollars each, on the first days of January, April, July and October, in each and every year of the said term.

ARTICLE FIFTEENTH.—The said party of the second part shall and will, at all times during the term aforesaid, and the continuance of this lease, keep an office in the City of Pittsburgh, which shall be open at all reasonable hours and times, for the transaction of the business of the said demised railways; and shall reserve and furnish, in the said office, free of charge, two suitable and convenient rooms, for the use of the President and Secretary, and Board of Directors, of the said party of the first part. And the said party of the second part shall, at all times during the said term, keep at the said office in the City of Pittsburgh, full, true and just accounts of any and all moneys received, and business done, upon the said demised railways and premises, and of all moneys paid, laid out and expended, and liabilities incurred in connection with the same; and also full statistical accounts, similar to those now kept by the said party of the first part, in or under the direction of a certain accounting department, designated as a Bureau of Statistics, at the office of the said party of the first part, in the City of Pittsburgh. The accounts to be kept by the said party of the second part, as above provided, and any and all accounts which shall or may be kept by the said party of the second part, in relation to the said demised railways, or the business of the same, shall at all reasonable hours and times, during the continuance of this lease, be open to the inspection and examination of the President of the said party of the first part, and of such other person or persons as the said party of the first part shall, from time to time, or at any time, by resolution of its Board of Directors, appoint to examine the same ; and the said party of the second part shall, at its own proper cost and expense, annually, to wit : on or before the first day of April, in each year, during the continuance of this lease, furnish to the party of the first part a detailed statement, duly authenticated, of the earnings, income and receipts arising from the said demised railways and premises, during the year ending with the 31st day of December last preceding the said first day of April ; and, also, a detailed statement, similarly authenticated, of all expenditures made by the said party of the second part, upon the said demised railways and premises, in the repair, replacement, renewal, improvement and equipment thereof, which statement shall specify the purposes of any and all such expenditures.

And it is hereby further agreed and declared that the party of the second part shall, at its own proper cost and expense, from time to time, and whenever necessary for the use of the party of the first part, make out and finish, to the said party of the first part, any and all reports and statements which the said party of the first part is now or may be hereafter required to make or file, under or by virtue of any law of either of the States of Pennsylvania, Ohio, Indiana or Illinois, now existing, or which may hereafter be enacted, or under any other lawful and competent authority.

ARTICLE SIXTEENTH.—The party of the first part hereby agrees, that, for the purpose of enabling the party of the second part to meet the obligations of the party of the first part to the public, by making from time to time such improvements upon and additions to the said Pittsburgh, Fort Wayne and Chicago

Railway, in the extension of facilities for increased business, by additional tracks and depots, shops and equipments, and the substitution of stone or iron bridges for wooden bridges, or steel rails for iron rails, the party of the first part will issue, from time to time, a special stock which shall bear such name as shall be hereafter agreed upon, or bonds, or other securities, which shall be issued in such form as may from time to time be found to be most available with respect to economy of interest and negotiability, and shall be consistent with the legal powers of the party of the first part and the rights secured by these presents; which special stock, or bonds, or other securities, shall be issued on the conditions following: The said party of the second part shall guarantee the payment, semi-annually or quarterly, thereon, of such rate of interest as may be agreed upon between the parties hereto, to be paid by the said party of the second part to the holders thereof, without deduction from the rent hereinbefore reserved; and the said special stock, or bonds, or other securities, shall be issued only in respect to improvements of and additions to the said railway which, and estimates and specifications of which, shall have been submitted to and approved by the said party of the first part, in writing; and all such improvements or additions shall be made in such manner as shall be approved by the said party of the first part. The party of the first part shall not at any time, during the term aforesaid and the continuance of this lease, make or issue any bond or obligation, in addition to the bonds hereinbefore specified, except subject to this case, without the consent in writing of the said party of the second part first had and obtained thereunto.

ARTICLE SEVENTEENTH.—Possession of the said demised railways and premises shall be given by the party of the first part to the party of the second part, on the first day of July, in the year one thousand eight hundred and sixty-nine; and upon the delivery of such possession the said party of the first part shall transfer and deliver to the said party of the second part, for use upon the said demised railways and premises, all machinery, tools, implements, furniture, fuel, material and other railroad supplies belonging to the said party of the first part, which shall have been procured for the use of the said railways, or either of them, and shall then remain on hand; and the said party of the first part shall settle, pay and discharge all wages, salaries and debts and liabilities incurred in operating the said demised railways, or either or any part of either thereof, up to that time, or for construction done or equipment received up to that time, and all other debts and liabilities which shall have matured, or been incurred, prior to the said first day of July, except those hereinbefore agreed to be paid or provided for by the said party of the second part; provided, and it is hereby agreed, that the said party of the second part shall and will, at the end of the term aforesaid, or other sooner determination of this lease, transfer and deliver in return, to the said party of the first part, machinery, tools, implements, fuel, materials and other railroad supplies equal in value to those delivered to it as aforesaid.

ARTICLE EIGHTEENTH.—The said party of the second part shall, and will, at all times during the term aforesaid, bear, pay and discharge, at its own proper cost and expense, and without deduction from the rent herein before reserved, any and all expenses, costs, damages, liabilities, claims and demands whatsoever, which shall or may arise out of the possession, management or operation of the said demised railways and premises, or of either, or of any of either thereof, or out of the business of the same ; and shall and will at all times during the term aforesaid, hold, save, and keep, harmless and indemnified, the said party of the first part of, from and against any and all expenses of operating the railways and premises hereby demised, and all damages, liabilities, actions and causes of action, suits, claims and demands for injuries to persons or property, or for causing the death of any person or thing through accident, neglect or default, during said term, or for breach of contract, or wrong done or suffered by the said party of the second part in the refusal to transport, or negligence in transporting any person, property, or thing, or by the loss, conversion, or non-delivery of any property which the said party of the second part shall have agreed, or undertaken, or be bound to transport over the said railways, or either or any part of either thereof, or which the said party of the first part, as the owner of the said railways hereby demised, or either or any part of either thereof, is or shall be under any legal obligation, by contract, public duty, or otherwise, to transport thereon ; and of, from and against any and all costs, damages, liabilities, actions and causes of action, suits, claims and demands whatsoever, which shall or may arise out of or in respect to the management, operation, or business of the said demised railways and premises, or either or any part of either thereof, during the term aforesaid ; and the said party of the second part shall and will defend all suits and claims which shall or may be brought against the party of the first part during the said term, in respect to any matter or thing arising out of the management or operation of the said demised railways or either or any part of either thereof, and indemnify and save harmless the party of the first part of, from and against any and all matters and things whatsoever, existing or to arise, which might or could be a charge upon or operate to reduce the rent hereinbefore reserved, or the fund aforesaid, to be applicable to the payment of dividends on the stock of the party of the first part, excepting only the debts and libalities incurred in operating the said demised railways, or either or any part of either thereof, prior to the first day of July, in the year 1869, which are mentioned in Article Seventeenth of these presents.

And it is hereby further declared and agreed that all the provisions of this article, in respect to indemnifying and saving harmless the party of the first part, shall apply to the railways and premises of which leases or operating contracts are hereby assigned, and each thereof, and the business of the same, during the terms of such leases or operating contracts, respectively, in the same manner and to the same extent as if the said railways of which leases or operating contracts were hereinbefore assigned were portions of the premises demised by these presents.

ARTICLE NINETEENTH.—In case the said party of the second part, its successors or assigns, shall at any time or times hereafter, during the term aforesaid, fail or omit to pay the rent hereinbefore mentioned or provided to be paid by the said party of the second part, or any part of such rent, when the same shall become payable as hereinbefore specified, or in case the said party of the second part, its successors or assigns, shall fail or omit to keep and perform the covenants and agreements herein contained, or any of them, and shall continue in default in respect to the performance of such covenant or agreement for the period of ninety days, then, and in either and every such case it shall be lawful for the said party of the first part, its successors or assigns, at its or their own option, to enter into and upon the railways and premises hereinbefore demised, and any and every part thereof, and remove all persons therefrom; and from thenceforth the said demised railways and premises, with the equipments and appurtenances thereof, and all additions and improvements which shall have been made to the same, to have, hold, possess and enjoy, as of the first or former estate of the said party of the first part in the said demised premises; and upon such entry for non-payment of rent or breach, or non-performance of any covenant or agreement herein contained, to be by the said party of the second part observed or performed, all the estate, right, title, interest, property, possession, claim and demand whatsoever of the said party of the second part, its successors or assigns, in or to the said demised railways and premises, or either or any part of either thereof, as well as all the right, title and interest of the said party of the second part, its successors or assigns, in, under or by virtue of the leases, contracts and agreements, or either of them, hereinbefore assigned or transferred to the said party of the second part, or asssumed by it, shall wholly and absolutely cease, determine and become void, anything hereinbefore contained to the contrary in any wise notwithstanding; but in case of re-entry, as aforesaid, the rent provided in Article First of these presents, and the several instalments thereof, shall be apportioned from the times of the last preceding payments of such instalments up to the time of such re-entry, and such portion thereof as would have been payable in respect to the intervening time, if the whole period in respect to which such instalments were payable had elapsed, shall be deemed and taken to be due and payable, and the same shall be paid by the said party of the second part; and it is further declared and agreed that such re-entry shall not waive or prejudice any claim or right of the party of the first part to or for damages against the party of the second part on account of such non-payment of rent, or non-performance or breach of the terms of this lease, and all such claims and rights are hereby expressly preserved to the said party of the first part.

ARTICLE TWENTIETH.—The said party of the second part hereby covenants, promises and agrees to and with the said party of the second part, that at the end of the said term, or other sooner determination of this lease, the said party of the second part shall re-deliver and surrender up to the said party of the first part, its successors or assigns, the said demised railways and premises, in at least as good order and condition as the same shall be delivered to the said party of the second part under this lease, and with such additions, betterments and improvements as shall have been made thereto, and also all the rolling stock, equipment and other property delivered under this lease, in as good order and condition as reasonable use and wear thereof, proper repairs and replacements thereof being made from time to time, will permit, or rolling stock, equipment and other similar property equal in value thereto, and also all additional rolling stock or equipment which shall be acquired or provided for use upon the said railways and premises, or any of them, or upon any of the railways, held by the party of the first part, under leases or operating contracts, which are assigned by these presents.

ARTICLE TWENTY-FIRST.—And the said party of the first part hereby covenants, promises and agrees to and with the said party of the second part, its successors and assigns, that the said party of the first part, its successors and assigns, shall and will, at any time or times hereafter, and whenever thereunto requested by the said party of the second part, its successors or assigns, execute, acknowledge and deliver to said party of the second part, its successors or assigns, at the proper cost and expense of the said party of the second part, its successors or assigns, any and all such other or further instruments and assurances in the law for the better demising and leasing of the said railways and premises to the said party of the second part, its successors and assigns, upon and subject to all and singular the rents, covenants, agreements and conditions hereinbefore reserved and mentioned, as by the said party of the second part, its successors or assigns, or by its or their counsel learned in the law, shall be reasonably advised, devised or required ; and the said party of the second part covenants, promises and agrees to and with the said party of the first part, its successors and assigns, that the said party of the second part, its successors or assigns, shall and will, at any time or times hereafter, and whenever thereunto requested by the said party of the first part, its successors or assigns, execute, acknowledge and deliver to the said party of the first part, its successors or assigns, any and all instruments for the more effectually assuring unto the said party of the first part, its successors and assigns, the payment of the rent hereinbefore reserved, or agreed to be paid, and the performance of the promises and agreements hereinbefore set forth on the part and behalf of the said party of the second part, to be performed, as by the said party of the first part, its successors or assigns, or by its or their counsel learned in the law, shall be reasonably advised, devised or required.

ARTICLE TWENTY-SECOND.—It is hereby expressly declared and agreed by and between the parties hereto, that these presents, and all the articles, covenants, agreements, terms and conditions thereof, shall take effect on the first day of July, 1869, and the same shall be binding upon the said parties hereto respectively, and their respective successors and assigns.

IN WITNESS WHEREOF, the parties hereto have caused their respective corporate seals to be hereunto affixed, and the same to be attested by the signatures of their respective Presidents and Secretaries, the day and year first above written.

| Seal P., F. W. & C. Railway Co. | G. W. CASS, *President P., F. W. & C. R'y. Co.* F. M. HUTCHINSON, *Secretary P., F. W. & O. R'y Co.* |

Sealed and delivered
  in the presence of—

THOS. D. MESSLER,
J. N. McCULLOUGH.

| Seal Penna. R. R. Company. | J. EDGAR THOMSON, *President Penn'a. R. R. Co.* JOS. LESLEY, *Secretary Penn'a. R. R. Co.* |

Sealed and delivered
  in the presence of—

H. J. LOMBAERT,
WM. J. HOWARD.

258

BONDHOLDERS' MEETING, TO ACT ON THE FORE-GOING LEASE.

Notice to the holders of the First and Second Mortgage Bonds of the Pittsburgh, Fort Wayne and Chicago Railway Co.:

In pursuance of the authority vested in the Trustees under the respective deeds of trust or mortgage securing the payment of the First and Second Mortgage Bonds, respectively, of the Pittsburgh, Fort Wayne and Chicago Railway Company, and in conformity with the By-law in relation to the meetings of said bondholders, adopted April 7, 1864, which provides that in the absence from the country of either of the Trustees, meetings of the bondholders may be called by the other Trustee, the under-signed Trustee under the said deeds, his associate Trustee being now absent from the country, hereby calls a meeting of the holders of the said First Mortgage Bonds, and also a meeting of the holders of the said Second Mortgage Bonds, to be held at the office of the said Company in the City of Pittsburgh, on the twenty-fourth day of June, 1869, at 12 o'clock, noon, of that day, for the purpose of considering and acting upon any and all such questions as may arise in reference to the lease of the rail-ways of the said Company to the Pennsylvania Railroad Company, or in reference to the conversion of the present stock of the said Company into a guaranteed stock of a larger aggregate, upon which dividends, at the rate of seven per centum per annum, payable quarterly out of the rental reserved in the said lease, shall be paid, and also for the purpose of considering and acting upon any and all other matters which may come before the said meetings or either of them.

S. J. TILDEN,
Trustee.

ACTION OF THE DIRECTORS

AND OF THE

STOCK AND BONDHOLDERS

IN ACCEPTANCE, APPROVAL & RATIFICATION

OF THE

FOREGOING LEASE OF THE COMPANY'S RAILROAD.

DIRECTORS' MEETING.

TUESDAY, June 22d, 1869.

The Board of Directors of the Pittsburgh, Fort Wayne and Chicago Railway Company met this day, at 12 o'clock, noon, at the principal office of the Company, in the City of Pittsburgh, pursuant to adjournment.

Present—GEORGE W. CASS, President; J. F. D. LANIER, S. J. TILDEN, LOUIS H. MEYER, SPRINGER HARBAUGH, PLINY HOAGLAND, JOHN L. DAWSON.

The President took the chair. The minutes of the last preceeding meeting (June 4th, 1869,) were read and approved; and there being no further business before the meeting, on motion of SPRINGER HARBAUGH, Esq., the Board adjourned to meet at the same place on Friday, the 25th inst., at 11 o'clock A. M.

FRIDAY, June 25, 1869.

The Board of Directors of the Pittsburgh, Fort Wayne and Chicago Railway Company met this day, at 11 o'clock, A. M., at the principal office of the Company, at the City of Pittsburgh, pursuant to the adjournment of the 23d inst.

Present—GEORGE W. CASS, President; J. F. D. LANIER, S. J. TILDEN, LOUIS H. MEYER, SPRINGER HARBAUGH, PLINY HOAGLAND, JOHN SHERMAN, JOHN L. DAWSON, KENT JARVIS.

The President took the chair and reported to the Board that since their last meeting he had caused the lease of the Pittsburgh, Fort Wayne and Chicago Railway and the other roads owned or leased by this Company, which connect therewith, to be duly executed, under date of June 7th, 1869; that the corporate seal had been thereto affixed, and the same attested by his own signature as President, and that of F. M. Hutchinson as Secretary.

Whereupon the following preamble and resolution were offered by Hon. JOHN SHERMAN, viz. :

Whereas, it is announced by the President of this Company to the Board of Directors, that the lease bearing date on the seventh day of June instant, and a copy whereof is set forth at length in the minutes of the meeting of this Board held on the 4th instant, has been duly executed on behalf of this Company, in pursuance of the resolutions of the Board passed on the said 4th instant; and whereas, at meetings of the stockholders and bondholders of this Company, held yesterday and to-day, at this city, the said lease to the Pennsylvania Railroad Company of the said railways has been approved and ratified by a large majority in interest of the said stockholders, and of each class of the bondholders; therefore,

Resolved, That the action of the President and Secretary in executing the said lease to the Pennsylvania Railroad Company be and the same is hereby approved, and that the President be and he is hereby authorized and directed to deliver to the said Pennsylvania Railroad Company, in conformity with the terms of the said lease, the railways and property therein embraced, to be held subject to and upon the agreements, terms and conditions of the said lease.

Which preamble and resolution were, on motion, unanimously adopted.

Mr. SHERMAN then offered the following further preamble and resolution, which were also, on motion, unanimously adopted, viz. :

Whereas, the railways of this Company shall, on the first day of July, prox., under and by virtue of the lease to the Pennsylvania Railroad Company pass under the management of the said Pennsylvania Railroad Company. which will make inoperative many of the By-Laws of this Company, and render superfluous the organization and general rules for the business of operating the railway; therefore,

Resolved, That from and after the first day of July, 1869, the organization for conducting the business of operating the railways of this Company, and the general rules in relation to such business, shall be, and the same are hereby declared to be suspended until the further orders of this Board, and from thenceforth until such further orders, any and all persons who may be employed by this Company, shall be subject to the direct orders of the President.

Resolved, That the President be and he is hereby authorized and directed to cause a revision of the By-Laws to be prepared, suited to the new condition upon which the Company is about to enter, and that he submit the same to this Board.

Resolved, That there be and is hereby constituted an Executive Committee of this Board, consisting of the President, who shall *ex officio* be Chairman, and three other members of the Board, which Committee shall have, when the Board is not in session, all the powers of the Board which can be lawfully delegated to or exercised by them. Such Committee shall keep a full record of all their proceedings, and submit the same to the Board from time to time as the latter may be convened.

Resolved, That the President of the Company, with James F. D. Lanier, Samuel J. Tilden and Louis H. Meyer, Esqrs., be and they are hereby constituted such Executive Committee until otherwise ordered by this Board.

Resolved, That until the further orders of the Board the monthly payments to be made by the Pennsylvania Railroad Company, under the lease mentioned in the foregoing resolutions, shall be made directly to Messrs. Winslow, Lanier & Co., of the City of New York, the Transfer Agents of this Company in that city, who will notify the Treasurer of each payment of said rent as the same shall be made.

Resolved, That the sum of $1,380,000, rental, reserved under the said lease as a dividend fund for the benefit of the stockholders, shall be distributed to the stockholders in quarterly dividends, payable on the first Tuesday following the first Monday in each of the months of January, April, July and October in each year, commencing with the dividend to become due in October, 1869, and that the transfer books of the Company shall be closed for two weeks previously to the payment of each dividend unless hereafter otherwise ordered by this Board.

J. F. D. LANIER, Esq., offered the following preamble and resolution, which, on motion, were adopted.

Whereas, the Company has for several years past anticipated the payment of the interest on its Third Mortgage or Income Bonds, by paying the same on the first days of January and July instead of the first days of April and October in each year.

But *whereas*, under the aforesaid lease the moneys applicable to the payment of such interest on the said Third Mortgage or Income Bonds, is not payable until the 28th days of March and September in each year ; therefore,

Resolved, That the interest on the said Third Mortgage or Income Bonds be hereafter paid on the days provided for the payment thereof, in the agreement of April 8th, 1864, made between this Company and the Trustees under the mortgage securing the payment of said bonds, to wit: on the the first days of April and October in each year, and that, as a concession to the said bondholders, the three months interest on the interest, which will accrue on said bonds up to the first day of October next, be paid on that day.

J. F. D. LANIER, Esq., also offered the following resolution, which, on motion, was unanimously adopted, viz. :

Resolved, That the regular quarterly dividend (No. 21) of two and one-half (2½) per cent., free of Government tax, upon the capital stock of the Company be and the same is hereby declared, for the quarter ending on the 30th day of June, instant, and that the same be payable on and after Friday, the sixteenth day of July, proximo, at the office of Winslow, Lanier & Co., No. 27 Pine street, in the City of New York, to the stockholders of this Company registered in New York, and at the office of the Treasurer in Pittsburgh to those registered at the last mentioned place, and that the transfer books close on Tuesday, the 6th day of July, 1869, at 2 o'clock P. M., and reopen on Thursday, the 22d day of July, 1869, at 10 o'clock A. M.

    *   *   *   *   *   *   *

SPRINGER HARBAUGH, Esq., offered the following preamble and resolution, which, on motion, were adopted, viz. :

Whereas, at the stockholders' meeting, held yesterday, the following resolution was adopted :

" *Resolved*, That as inasmuch as in and by the provisions of the lease of the railway of this Company to the Pennsylvania Railroad Company, after satisfying the liabilities of the Company, for interest and the sinking funds, a perpetual dividend fund is provided adequate to pay twelve per cent. upon the existing stock of this Company, free and clear of all taxes which may operate as a deduction from the said dividends, it is expedient that a guaranteed stock, entitled to dividends at the rate of seven per cent. per annum, payable quarterly, in such form, and with such guarantees as the Board of Directors may prescribe, and of such aggregate amount as the annual rental of one million three hundred and eighty thousand dollars shall suffice to pay dividends upon, at the aforesaid rate of seven per cent. per annum, shall be created ; and that the same shall be issued in substitution of the now existing stock in such manner and on such terms as the Board of Directors may provide, and that the Board of Directors and such committee or officers as they may designate, are hereby vested with all powers which the stockholders can confer, and which may be necessary or proper to carry this substitution into complete effect."

And *whereas*, at meeting of the First and Second Mortgage Bondholders, respectively, duly convened and held at this City, yesterday, and to-day, each of the said classes of bondholders, did by a large majority, in interest, consent to the creation and issue of the said capital stock, to be issued in substitution for the existing stock, to such aggregate amount, that the sum of one million three hundred and eighty thousand dollars, rental reserved, as a dividend fund under the said Lease to the Pennsylvania Railroad Company, should be equal to not less than seven per cent. per annum thereto, therefore be it

RESOLVED, That the Executive Committee be and is hereby vested with authority to do every act and thing which may be necessary or proper to carry the aforesaid resolutions of the stockholders into full and complete effect, in such manner as in their judgment shall be most advisable.

JNO. L. DAWSON, Esq., offered the following resolution, which, on motion, was adopted, viz. :

Resolved, That the Executive Committee be and is hereby directed and empowered to make such contract with the Grand Rapids and Indiana Railroad Company, or such Company as may become the owner of said Grand Rapids and Indiana Railroad, as is contemplated in the lease to the Pennsylvania Railroad Company, and as authorized by resolution of the stock and bond holders, at the adjourned meeting held yesterday ; *Provided*, there shall be a stipulation in said contract to require the. first moneys arising from the sale of any and all bonds which may be guaranteed by this Company, to be expended in the construction of said road northward continuously from Fort Wayne, Indiana.

       *   *   *   *   *   *   *

ADJOURNED ANNUAL MEETING

OF THE

STOCK AND BONDHOLDERS

OF THE

PITTSBURGH, FORT WAYNE AND CHICAGO RAIL-WAY COMPANY.

PITTSBURGH, FORT WAYNE AND CHICAGO RAILWAY CO.,
Office of the Secretary,
PITTSBURGH, June 5, 1869.

By virtue of authority conferred by resolution of the Stock and Bond-holders of the Pittsburgh, Fort Wayne and Chicago Railway Company, at the annual meeting, held at the office of the Company, in this city, March 17, A. D. 1869, an adjourned meeting of said annual meeting, will be held at the general office of the Company, in the City of Pittsburgh, at 10 o'clock, A. M., of June 24th, inst., for the purpose of considering and acting upon a Lease for a period of nine hundred and ninety-nine years, of the railway and property of this Company, to the Pennsylvania Railroad Company, and to act upon such other business as may come before said adjourned meeting. The books for the transfer of stock and bonds of the Pittsburgh, Fort Wayne and Chicago Railway Company will close at 2 P. M., on Monday the 14th of June, at the Agency in New York, Winslow, Lanier & Co., 27 Pine Street, and at the Office in Pittsburgh, and will re-open on the 25th of June.

By order of the President,

F. M. HUTCHINSON, *Secretary.*

PITTSBURGH, June 24th, 1869.

A meeting of the Stockholders and Bondholders of the Pittsburgh, Fort Wayne and Chicago Railway Company, was held this day, commencing at 10 o'clock, A. M., at the principal office of the Company, in the City of Pittsburgh, pursuant to an adjournment, and the foregoing notice, which was duly published in conformity with the resolution passed 17th March, 1869.

The meeting was organized by JAMES S. CRAFT, Esq., of

Pittsburgh, resuming the Chair ; F. M. Hutchinson being Secretary.

GEORGE W. CASS, the President of the Company, presented the form of Lease, bearing date June 7, 1869, for a period of nine hundred and ninety-nine (999) years, of the Railways of this Company to the Pennsylvania Railroad Company, which was read by the Secretary and is as follows :

(For Lease, see page 225.)

And after discussion of the same

SAMUEL J. TILDEN, Esq., presented and moved the adoption of the resolutions hereinafter set forth, and marked Nos. 1, 2 and 3, which resolutions were read by the Secretary, as follows, viz. :

No. 1.

Resolved, That we, the stockholders and bondholders of the Pittsburgh, Fort Wayne and Chicago Railway Company, hereby assent to and approve of the lease of the railways of the said Company to the Pennsylvania Railroad Company, for the period of nine hundred and ninety-nine years, upon the terms and conditions, and for the annual rental reserved and mentioned in a certain Indenture of Lease thereof, bearing date on the seventh day of June, 1869, executed by, or on behalf of, the said Pennsylvania Railroad Company, and now just submitted to us ; and that we hereby assent to and approve of the execution and delivery of such lease on behalf of the said Pittsburgh, Fort Wayne and Chicago Railway Company: provided, however, that the assent and approval shall not be construed to waive, alter, or impair the existing liens of the respective deeds of trust or mortgage, by which the payment of the respective bonds is secured, but that this assent and approval is hereby expressly declared, and is given, upon the condition, that the said lease shall be, and shall ever hereafter be, deemed and taken to be subordinate to the aforesaid liens.

No. 2.

Resolved, That the assent of the stockholders is hereby given to the making by the Board of Directors of this Company, of such contract as is contemplated in Article Thirteenth of the lease this day submitted, upon the terms and conditions and upon the approval and assumption by the lessee as in the said article contemplated.

No. 3.

Resolved, That inasmuch as in and by the provisions of the lease of the railway of this Company to the Pennsylvania Railroad Company, after satisfying the liabilities of the Company for interest and sinking funds, a perpetual dividend fund is provided adequate to pay twelve per cent. upon the existing stock of this Company, free and clear of all taxes which may operate as a deduction from the said dividends, it is expedient that a guaranteed stock entitled to dividends at the rate of seven per cent. per annum, payable quarterly in such form and with such guaranties as the Board of Directors may prescribe, and of such aggregate amount as the annual rental of one million three hundred and eighty thousand dollars shall suffice to pay dividends upon at the aforesaid rate of seven per cent. per annum shall be created, and that the same shall be issued in substitution of the now existing stock, in such manner and upon such terms as the Board of Directors may provide, and that the Board of Directors and such committee or officers as they may designate are hereby vested with all powers which the stockholders can confer, and which may be necessary or proper to carry this substitution into complete effect.

The question being, what action will the meeting take upon the above resolutions presented by Mr. Tilden, George W. Cass, the President, moved that they be submitted separately to the vote of the stock and bond holders, and that the stock and bond holders vote thereon by ballot, which was agreed to; whereupon, the Chairman appointed Messrs. J. T. Brooks, James P. Farley and Wm. Leaf tellers, to receive, count and make record and return of the vote cast; and the stock and bond holders thereupon proceeded to vote upon such resolutions separately, and the voting being concluded, George W. Cass moved that the meeting do now adjourn for the purpose of enabling the tellers to count and make a report of the votes cast, and that it stand adjourned until to-morrow morning at 10 o'clock, which was agreed to; whereupon the meeting adjourned accordingly.

FRIDAY MORNING, June 25, '69.

Agreeably to the adjournment of yesterday, the stockholders and bondholders of the Pittsburgh, Fort Wayne and Chicago

Railway Company met at the general office of the Company, in the City of Pittsburgh, at 10 o'clock this day, J. S. Craft being Chairman and F. M. Hutchinson, Secretary.

J. F. D. Lanier, Esq., presented and read to the meeting the following return of the tellers appointed at yesterday's meeting to receive and count the votes to be cast upon the adoption or rejection of resolutions submitted yesterday by Mr. Tilden, and numbered one, two, and three, respectively, which return was as follows :

The subscribers having been chosen and duly qualified to act as tellers of election at the meeting of the stock and bond holders of the Pittsburgh, Fort Wayne and Chicago Railway Company, held at their office in the City of Pittsburgh, on the 24th day of June, 1869, having fully discharged their duties in the premises, submit the following report : At said meeting three separate and independent resolutions were submitted to the stock and bond holders, numbered consecutively one, two and three. The first of said resolutions expressing the concurrence of the stock and bond holders of said Railway Company in the lease of the Railway and appurtenances of said Company to the Pennsylvania Railroad Company.

The second expressing the assent of the stockholders to the making of a certain contract by the Directors of said Company, with the Grand Rapids and Indiana Railroad Company, and to the assumption of said contract by said Pennsylvania Railroad Company.

The third declaring it to be expedient that a guaranteed stock entitled to dividends at the rate of seven per cent. per annum, be created and issued in substitution of the existing stock of said Pittsburgh, Fort Wayne and Chicago Railway Company, which resolutions in full are a part of the proceedings of this meeting.

We do further find that the full number of votes cast at said election was one hundred and eighty-six thousand nine hundred and eighteen (186,918), composed as follows :

1st Mortgage Bonds.	2d Mortgage Bonds.	3d Mortgage Bonds.	Equipment Bonds.	Stocks.
42,610	38,280	14,870	8,800	82,358

In favor of said first resolution, one hundred and seventy-nine thousand seven hundred and one (179,701); against said first resolution, seven thousand two hundred and seventeen (7,217); in favor of said second resolution, one hundred and eighty-six thousand nine hundred and eighteen (186,918); against said second resolution, none.

In favor of said third resolution, one hundred and sixty-two thousand six hundred and thirty-six (162,636); against said third resolution, twenty-four thousand two hundred and eighty-two (24,282), all of which is respectfully submitted.

> J. T. BROOKS,
> J. P. FARLEY, } *Tellers of Election.*
> WM. LEAF.

Which said report was ordered to be entered at length, upon the minutes.

A majority of the votes having been cast in the affirmative, on each of the said resolutions, the Chairman declared each of them adopted.

George W. Cass, Esq., then moved that when this meeting adjourn, it will adjourn to meet at the call of the President of the Company, upon ten days' notice; which was agreed to, and the meeting adjourned accordingly.

> F. M. HUTCHINSON,
> *Secretary.*

REVIEW OF THE HISTORY

AND

MANAGEMENT OF THE RAILWAY.

(Published by the N. Y. Times, in July, 1869.)

An example of able, successful, and honest railway management presents itself to notice and discussion, as a matter of the
first importance to all readers interested, as stock and bond
holders or otherwise, in the railway system of the United States,
by reason of the recent lease of one of our great western lines, to
the Pennsylvania Railroad Company.

In 1859, the Pittsburg, Fort Wayne and Chicago Railroad,
in common with most other lines, was overwhelmed in the financial revulsion which had swept with resistless force over
the whole country. The road had been just opened to Chicago.
The line was originally undertaken by three companies, none of
which possessed means at all adequate to the construction of their
several links. The road, when opened, was hardly more than
half completed. Its earnings not equaling one-quarter their
present amount, were wholly insufficient to meet current expenses
and the interest on its funded debt.

Default, by necessary consequence, was made on all classes
of its securities. Bankruptcy stared the concern full in the face,
threatening the loss of nearly the whole amount invested.

In this crisis, a meeting of its creditors, chiefly First Mortgage
Bondholders, was called at the office of Winslow, Lanier & Co.
to consider what was to be done. This class of creditors, of
course, had the precedence. If they insisted upon the letter of
the law, they would inevitably cut off all subsequent parties in
interest, who represented an amount of capital invested in the
road twice greater. After much deliberation it was decided to
raise a Committee, to be invested with full power, and if possible
save the interests of all.

This Committee consisted of Mr. J. F. D. Lanier, who was
appointed by the creditors its chairman ; Mr. Samuel J. Tilden ;
Mr. Louis H. Meyer ; Mr. J. Edgar Thomson, President Pennsylvania Railroad, and Mr. Samuel Hanna, of Fort Wayne.

To give some idea of the chaos existing in the affairs of the Company, we may state that there were outstanding at the time nine different classes of bonds, secured, in one way or another, upon the different portions of the road ; two classes secured by real estate belonging to the Company, and several issued in the funding of coupons. Upon all these interest for several years, amounting to many millions of dollars, was overdue. The principal sums of several of the first mortgages were speedily to mature. The Company also owed more than $2,000,000 of floating debt, portions of it in the form of judgments recovered in the State Courts. The road was in extremely bad condition, and required the expenditure of a large sum to enable it to conduct its business with any degree of economy or dispatch.

Such was the condition of affairs when the Committee commenced work. The value of the securities of the Company was merely nominal. Its stock would not sell for five cents on the dollar. Each class of creditors was striving to gain some advantage at the expense of the others. The first step of the Committee, consequently, was to put the property beyond the reach of individuals and in the custody of the Courts. An order for this purpose was obtained in the United States District Court for the Northern District of Ohio, on the 17th of January, 1860, and Mr. Wm. B. Ogden was appointed Receiver.

The Committee set out with the determination of preserving, if possible, the rights of all the parties in interest, not alone those of the First Mortgage Bondholders. It was hoped that when the property was put beyond the reach of individual creditors, an arrangement might be effected, and the rights of the various parties preserved in the relations they had previously maintained.

But such an adjustment required the assent of each creditor and stockholder. This, in the multiplicity and conflict of interests, it was found impossible to obtain. The next and only remaining course was to sell the road and property of the Company by an order of Court in behalf of the First Mortgages. Such sale would vest absolutely the title to the road in the hands of the purchasers, who would thus be in position to make such disposition of it as in their view equity and justice might demand. It would also enable them to apply the net earnings to the construction of a good road, without which the investment itself would be of no value.

With this purpose a full plan of re-organization, such as was finally adopted, was prepared and published, and brought as far as possible, to the attention of every party in interest. Decrees for sale had to be obtained in the Courts of the United States of four different States. The time required for this purpose was occupied by the committee in incessant efforts in removing one impediment after another thrown in the way by unfortunate and dissatisfied creditors, who were indifferent to the fate of the concern, provided they could get their pay. All difficulties were at last overcome, and on the 24th of October, 1861, the road and property was sold at auction and purchased by Mr. Lanier, in behalf of himself and his associates, for the sum of $2,000,000. The Courts, we are happy to say, facilitated legal proceedings, as far as this could be properly done. They had full confidence in the committee, and sympathized with the unfortunate creditors of the concern, and not as at the present day in our State, with bands of conspirators against the public welfare, who seek the control of great lines, with no other purpose but to plunder them. Eight years ago, measured by what has since transpired, was a golden age of judicial purity.

By the sale of the road a most important step was gained. The title to it vested absolutely in the purchasers. They could convey to whom, at what price and upon what terms they pleased. What followed was more a matter of detail, though involving great patience and labor. For the creation of a new Company, according to the original plan of re-organization, legislation had to be obtained in the States of Pennsylvania, Ohio, Indiana and Illinois. Such legislation was at last secured, a new Company formed, to which was conveyed the railroad and everything appertaining thereto, the committee receiving therefor, first, second and third mortgage bonds, in amounts sufficient to meet the sums due the different classes of creditors in the old Company, and also certificates of stock corresponding in amount to that outstanding in the old. First Mortgage Bonds, to the amount of $5,200,000, were issued to the First Mortgage Bondholders of the old Company, and of the several links of which its road was composed, and for accrued interest. The bondholders were also required to fund for two years, the interest accruing on the new bonds, so as to allow, for such a period, the application of the net earnings to construction.

The Second Mortgage Bondholders received, in the same manner and subject to similar conditions, Second Mortgage Bonds to the amount of $5,250,000. The unsecured creditors were paid off in Third Mortgage Bonds to the amount of $2,000,000. The shareholders received new certificates in exchange for the old. By these means each class of creditors without the abatement of a dollar, were fully and completely reinstated in the new company in the order they stood in the old. The proper transfers and exchanges were made, and on the 1st day of May, 1862, two years and six months after the road was placed in the hands of a receiver, and six months after the sale, the trust so long held and faithfully executed was brought to a virtual close, to the entire satisfaction of every party in interest in the road.

During the period of reorganization the road was operated under the general direction of the Committee by Mr. Geo. W. Cass, its former and subsequent President. His well known abilities as a railroad manager were never more conspicuously displayed than in this service. He had every difficulty to contend with—an impoverished and half completed road, clamorous creditors at every turn.

The Chairman of the Committee was not unfrequently called upon to advance, from his private funds, considerable sums in aid of the operations of the road. Such advances were, of course, repaid, but only with simple interest. The good name and financial strength of Mr. Lanier, joined to well known prudence and caution, tended to inspire great confidence in the action of the Committee, in which he justly exerted great influence. Mr. Thomson's position as chief of a great and successful enterprise enabled him to render very great aid to the Committee in the operations of the road. Indeed, it was through his instrumentality that the old Company was enabled to push its line through to Chicago. Mr. Tilden was the chief legal adviser of the Committee and Company throughout. He had charge of the proceedings not only for the winding up of the old, but for the formation of the new Company, and for the recent transfer of the road to the Pennsylvania Company, and drew up all the documents and guaranties relating to the same. The proper discharge of his duties involved the fate and security of the whole investment. Not a suggestion has been ever raised that they were not ably and faithfully performed. The directors of the Company, pending its reorganization, rendered valuable assistance. Many of them resided upon the line of the road, and were enabled to exert a salutary influence, not only among the creditors of the Company, but in securing the legislation required. But it is, perhaps, invidious to particularize when all worked faithfully and well. Not a dollar was ever paid to secure the legislation required for the formation of the new Company; not a dollar to buy off importunate or unreasonable creditors. The Committee never had a secret which they turned to account at the expense of the stock and bond holders. Their plans were prepared and published in the outset and scrupulously adhered to.

Soon after the new Company commenced operations it was seen the enterprise had passed its darkest days. For the year ending Dec. 31, 1862, the net earnings of the road equaled nearly $2,000,000, all of which were applied to construction.

The Committee was enabled to add largely to its available means by the sale of property purchased with the road, but not needed in its future operations, and which, in fact, they were not, by the terms of the trust, to account for to the new Company. The sums realized from these sources and paid over to the Company equaled about $600,000, of which some $400,000 was saved by a compromise which the Committee were enabled to make with European holders of bonds secured by real estate. All the advantages gained by such settlements were given to the new Company.

In 1863 the net earnings equaled nearly $3,000,000. These sums enabled the Company to place its road in first rate condition, and on the 1st day of April, 1864, it commenced the payment of dividends at the rate of 10 per cent. per annum, free of Government tax, in quarterly payments of $2\frac{1}{2}$ per cent. each. These were continued regularly to the 1st day of July, 1869, when the road was leased to the Pennsylvania Railroad Company for 999 years, at an annual rental of 12 per cent. on its share capital.

In this lease the Pennsylvania Company assumes every obligation or charge for which the Fort Wayne Company are, or may be, liable. It pays the sum of $19,000 annually for the maintenance of the organization of the former. It keeps up the annual contributions to the sinking fund. These contributions will, in twenty-six years, wholly pay off the bonded debt of the Fort Wayne Company, leaving the stockholders the sole owners of the road; and, in conclusion, it agrees to pay an annual rental of $1,380,000, a sum which equals 12 per cent. annually upon the stock, *free* of *Government tax*, or of any other charge. The terms of the lease also allow the Fort Wayne Company to increase its share capital *seventy-one and three-sevenths per cent.*, and to issue certificates for the whole capital, upon which, for the entire period of the lease, *seven per cent. a year, in quarterly payments of one and three-quarters per cent.*, free of Government tax, is to be paid. All these payments, as well as the accruing interest, is to be made directly to the agency of the Fort Wayne Company in New York. When we consider that the net earnings of the road largely exceed the rental paid, and that this rental is guaranteed by the most powerful and successful railroad corporation on this continent, and that the lease will inure even more to its advantage than to that of the lessors, in placing a common line under a common head and management, certainly it is not within the power of man to make a better security, or one in which trust funds can be more securely placed.

We have thus put on record a detailed statement of the resuscitation and success of a great enterprise, as an example of what has been and may be accomplished by upright, able and public-spirited men. In no country do railways bear a relation to the internal economy of a people so intimate as in ours. No investments, consequently, can be so productive as those made in good and well managed lines. There is no doubt that the gross earnings of the railroads of the Northern States equal fully 30 per cent. annually of their actual cost. One-third of this, at least, should be net; and we take pleasure in placing an illustration before our readers, where the best possible net result has not only been secured, but secured, as it should be, to those that are and have been the owners of the property.

FORM OF NEW CERTIFICATE OF STOCK.

PITTSBURGH, FORT WAYNE AND CHICAGO RAILWAY COMPANY.

No.   [Viguette.]   Shares.

18 Shares.

Registrar of Transfer.

Dividends payable on the Tuesday following the first Monday of Jany., April, July and October in each year.

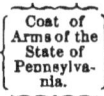

Coat of Arms of the State of Pennsylvania.

Coat of Arms of the State of Ohio

Coat of Arms of the State of Indiana.

Coat of Arms of the State of Illinois.

day of

The Third National Bank of the City of New York.

Countersigned and registered

By

This certifies that
the owner of
Shares of One Hundred Dollars each in the Capital Stock of the Pittsburgh, Fort Wayne and Chicago Railway Company, transferable only on the books of the Company, at their Agency in the City of New York, in person or by Attorney, and according to the regulations of the Company on the surrender of this Certificate.

This Certificate will not be valid until countersigned by Winslow, Lanier & Co., Transfer Agents in the City of New York, and also by the Third National Bank of the City of New York, Registrar of Transfers in the said City.

The holders of this stock are entitled to the benefits of the guaranties of the Pennsylvania Railroad Company, expressed in a lease by this Company to the Pennsylvania Railroad Company creating a dividend fund adequate to pay seven per cent. per annum upon the said stock, payable quarterly, free from income and other taxes, as specified in Articles First and Sixth, of said lease, extracts from which are endorsed thereon.

IN WITNESS WHEREOF, the said Company have caused their Corporate Seal to be affixed hereto, and this Certificate to be signed by their President and Secretary, at the Office of the Company in the City of Pittsburgh, Pennsylvania, this
day of   18

—————————Secretary. ——————President.

Countersigned this   day }
of   ,   18   }

Transfer Agents.

(Endorsed with extracts from Articles First and Sixth of Lease.)
(Also, form of Transfer.)

CONTRACT

WITH THE GRAND RAPIDS AND INDIANA RAILROAD COMPANY, ASSUMED BY THE PENNSYLVANIA RAILROAD COMPANY.

THIS INDENTURE made the thirtieth day of September, in the year one thousand eight hundred and sixty-nine, between the PITTSBURGH, FORT WAYNE AND CHICAGO RAILWAY COMPANY, duly formed and organized under the laws of the States of Pennsylvania, Ohio, Indiana and Illinois, party of the first part, and the GRAND RAPIDS AND INDIANA RAILROAD COMPANY, duly formed and organized under the laws of the States of Michigan and Indiana, party of the second part.

Whereas, the said party of the first part is the owner of a certain line of railway, commonly known as the Pittsburgh, Fort Wayne and Chicago Railway, which extends from the City of Pittsburgh, in the State of Pennsylvania, by way of Crestline, in the State of Ohio, to Fort Wayne, in the State of Indiana, and from thence to Chicago, in the State of Illinois.

And whereas, the said party of the second part is duly authorized by its charters to construct and operate a railroad, to be known as the Grand Rapids and Indiana Railroad, from the City of Fort Wayne aforesaid, northerly through the States of Indiana and Michigan, to a point on little Traverse Bay, in the County of Emmet and State of Michigan, being a distance of three hundred and twenty (320) miles, more or less; which railroad the said party of the second part is authorized by law to connect with the said Pittsburgh, Fort Wayne and Chicago Railway, at or near Fort Wayne, aforesaid.

And whereas, in order to procure the means for the construction of its railroad, the said party of the second part is about to issue a series of eight thousand bonds, of one thousand dollars each, amounting in the aggregate to eight millions of dollars; which bonds are to be payable on the first day of October, in the year one thousand eight hundred and ninety-nine, and bear interest from the first day of October, in the year one thousand eight hundred and sixty-nine, at the rate of seven per centum per annum, payable semi-annually, to wit, on the first days of January and July, as to one-half of the said series of bonds, and on the first days of April and October, as to the other half of the same, and the payment whereof is secured by a deed of trust or mortgage upon the said Grand Rapids and Indiana Railroad, with its equipments, appurtenances, and franchises.

And whereas, it has become necessary for the said party of the second part, in order to negotiate said bonds, to obtain the assistance, and the loan and security of the credit and responsibility of the Pittsburgh, Fort Wayne and Chicago Railway Company owning or operating the said Pittsburgh, Fort Wayne and Chicago Railway, with the road of which, as aforesaid, the said Grand Rapids and Indiana Railroad is to be connected at or near Fort Wayne, aforesaid.

And whereas, the said party of the second part heretofore applied to the said party of the first part to aid in the construction of the said Grand Rapids and Indiana Railroad, by guaranteeing, or becoming surety for, the payment of the principal and interest of a portion of the bonds to be issued as aforesaid by the said party of the second part, and thereby enable the said party of the second part to negotiate the same, in consideration that the said party of the second part would secure to the said party of the first part, and its assigns, any and all business which might be contributory, by the said Grand Rapids and Indiana Railroad to the said Pittsburgh, Fort Wayne and Chicago Railway, and thereupon a negotiation between the said Companies, in respect to such application, ensued.

And whereas, pending the said negotiation, the said party of the first part, by an Indenture of Lease, bearing date on the seventh day of June, 1869, demised and leased to the Pennsylvania Railroad Company, all and singular the said Pittsburgh, Fort Wayne and Chicago Railway, with its equipments and appurtenances, and the other premises described in the said Indenture of Lease, for the term of nine hundred and ninety nine years, upon the terms therein set forth, and at a certain fixed rental.

And whereas, for the purpose of securing, to the lessee under the said Indenture, the benefits to be derived from the construction of the said Grand Rapids and Indiana Railroad, and the connection thereof with the said Railway, the following provision was inserted in the said Indenture of Lease, to wit :

ARTICLE THIRTEENTH.— *Whereas,* a negotiation is now pending between the party of the first part and the Grand Rapids and Indiana Railroad Company for the purpose of securing to the party of the first part or its assigns, any and all such business as may be contributory by the said Grand Rapids and Indiana Railroad Company to the said party of the first part or its assigns, in the connection of the respective roads of said companies, and thereby the better to develop the business of each Company.

" *Now, therefore, this Indenture witnesseth,* that the said party of the first part has agreed, and does hereby promise and agree, to and with the said party of the second part, that the said party of the first part shall and will assign and transfer, to the said party of the second part, such contract, if any, as, shall or may be concluded, by the said party of the first part with the said Grand Rapids and Indiana Railroad Company, for the purposes aforesaid, in consideration that the said party of the second part shall assume and undertake the performance of such contract, and indemnify and save harmless the said party of the first part, of, from and against the same, and any and all claims, demands and liabilities, whatsoever, which can or may arise thereunder ; and the said party of the second part hereby agrees to assume such contract, and to indemnify and save harmless the said party of the first part from and against the same, and all claims and demands thereunder ; provided such contract or proposed contract shall not involve the guaranty of an amount exceeding four millions of dollars of first mortgage bonds of the said Grand Rapids and Indiana Railroad Company, the whole issue of which First Mortgage Bonds shall not exceed eight millions of dollars, and which said issue shall be further secured by a deed of trust of one million of acres of land, and that a majority in amount of the capital stock of the said Grand Rapids and Indiana Railroad Company shall be deposited in trust until the principal and interest of said guaranteed bonds shall be provided for, and the guaranty satisfied and canceled, and that the form of such contract or proposed contract shall be submitted to and approved by the Board of Directors of the party of the second part."

Now, THEREFORE, THIS INDENTURE WITNESSETH, that the said parties hereto, in consideration of the premises, and of the mutual covenants herein contained, have promised and agreed, and do hereby covenant, promise and agree, to and with each other, in the manner and form following, that is to say :

ARTICLE FIRST.—The said party of the first part hereby promises and agrees, to and with the said party of the second part, for the benefit of the person or persons who may hereafter become the holder or holders of the several bonds and coupons hereinafter mentioned, or any of them, and in order to enable the said party of the second part to negotiate said bonds and coupons, that in case the said party of the second part shall, at any time or times hereafter, be unable, or fail or omit, to pay, upon presentation at the maturity thereof and demand of payment, according to the terms thereof, any of the bonds aforesaid, designated as the First Mortgage Bonds of the said party of the second part, and included in the issue, or series thereof, numbered from one (1) to four thousand (4,000), inclusively, being one half of the entire issue of said bonds, or any of the coupons thereunto annexed or belonging ; and if such bond or coupon shall remain due and unpaid for the period of sixty days, after notice, in writing, to the party of the first part, or its assigns, of the due presentment thereof, the said party of the first part, or its assigns, shall and will, upon demand thereof, purchase, pay for, and take such over-due bond or coupon, from any holder or holders thereof, giving such notice, or his or their assigns—excepting only the Continental Improvement Company, a corporation of the State of Pennsylvania—at the par value of the same ; *Provided*, that there shall then be annexed to the said bond the memorandum of agreement, or certificate, hereinafter mentioned, indicating that such bond is entitled to the benefit of this agreement, signed by the President of the said party of the first part; *and provided further*, and it is hereby expressly declared and agreed, that each and every bond and coupon which may be purchased, as aforesaid, by the said party of the first part, or its assigns, shall continue in full force and validity after such purchase, and remain and be a debt, secured by mortgage, payable by and enforceable against, the said party of the second part, in the same manner and to the same extent as before such purchase.

And the said party of the first part hereby further promises and agrees, that it shall and will cause to be annexed to, and issued in connection with, each of the said four thousand bonds of the issue, numbered from 1 to 4,000 inclusively, a memorandum of agreement, or certificate, in the form following, to wit:

"GUARANTY OF ANNEXED BOND AND COUPONS, BY AGREEMENT FOR THE PURCHASE THEREOF AT PAR.

" Any holder of the annexed Bond, number   , or of any Coupon thereunto belonging—except the Continental Improvement Company—will, if such Bond or Coupon remain unpaid after maturity and demand of payment, become entitled, upon sixty days' notice, to have the same purchased at par, under and by virtue of a certain agreement, bearing date on the 30th day of September, 1869, made between the Pittsburgh, Fort Wayne and Chicago Railway Company, and the Grand Rapids and Indiana Railroad Company, and a certain other agreement, bearing the same date, made between the Pennsylvania Railroad Company, and the said Pittsburgh, Fort Wayne and Chicago Railway Company, in pursuance of Article Thirteenth of the Lease of the Pittsburgh, Fort Wayne and Chicago Railway, theretofore executed to the said last mentioned Company.

IN WITNESS WHEREOF, The said Pittsburgh, Fort Wayne and Chicago Railway Company has caused these presents to be signed by its President, the first day of October, 1869.

President Pittsburgh, Fort Wayne and Chicago Railway Company.

ARTICLE SECOND.—The said party of the second part has promised and agreed, and does hereby promise and agree, to and with the said party of the first part, as follows, to wit. :

1. That the said party of the second part, as an additional security for the payment of the principal and interest of the bonds

aforesaid, shall and will, before the issue of any of the said bonds, duly execute, acknowledge and deliver, on its own behalf, and cause to be duly executed, acknowledged and delivered, on behalf of the Continental Improvement Company aforesaid, (with which company the said party of the second part has entered into a certain contract, bearing date on the first day of May, 1869, relative to the construction of its Rail Road, a deed of trust or mortgage, in and by which the said party of the second part, and the said Continental Improvement Company, shall and will grant and convey to JOHN EDGAR THOMSON and GEORGE W. CASS, as trustees, all the right, title and interest which the said party of the second part and the said Continental Improvement Company now have, or either of them has, or shall or may at any time hereafter acquire, of, in, or to the whole or any part of, all those certain lands, situate in the State of Michigan, which were heretofore granted by the Congress of the United States to the State of Michigan, and by the said State to the said Grand Rapids and Indiana Railroad Company, to aid in the construction of its said railroad—which said lands are estimated, in and by the said grants thereof, to comprise in the aggregate one million and one hundred and sixty thousand (1,160,000) acres, more or less—to be held by the said trustees, and the survivor of them, and their and his successors and successor, in trust to secure the payment of the bonds aforesaid, and the several coupons, thereunto belonging, as the said bonds and coupons, respectively, shall become due, subject, however to a provision, to be inserted in said deed of trust or mortgage, whereby the said lands may be sold at a reasonable valuation, and the proceeds thereof applied as a sinking fund for the purchase of the said bonds, for the benefit of the said Continental Improvement Company.

And, as a further security for the payment of said bonds and coupons, and indemnity to the party of the first part hereto, and its assigns, against any and all loss or damage on account thereof, or of the aforesaid agreement for the purchase of the same, the said party of the second part shall and will, from time to time, transfer, or cause to be transferred and assigned, to John Edgar Thomson, Herman J. Lombaert and Edmund Smith, as trustees, with power to vote thereon at all elections for directors, and at all meetings of stockholders of said Company, such amount or amounts of the capital stock of the said party of the second part as shall at all times keep in the possession and under the control of the said trustees a majority of the entire capital stock of the said party of the second part which may be issued, to be held by the said trustees until the said four thousand bonds, numbered from 1 to 4,000, inclusive, shall be paid, or the memorandum of agreement, or certificate relative to the purchase thereof, hereinbefore mentioned, duly canceled; and in the event of the death, resignation, neglect, refusal, or incapacity to act of the said trustees, or either of them herein named, then the party of the second part hereto shall and will appoint such new trustee or trustees as may be nominated and designated by a majority of the Board of Directors of the Pennsylvania Railroad Company for the purpose of filling the vacancy so caused, and supplying the place of the said trustee so dying, resigning, neglecting, refusing or becoming incapable to act, and the said trustee or trustees so nominated and designated and appointed, shall take upon himself, or themselves, the same trusts, and have the same powers as herein mentioned, and which it is hereby agreed and declared shall extend to and be performed by such newly nominated and appointed trustee or trustees. And the like nomination, designation and appointment shall and may be made and carried into effect in like manner and as often, from time to time, as there may be occasion therefor, and with the same effect as before mentioned.

And it is hereby further agreed, that the said party of the second part shall annex to, and cause to be duly executed and issued in connection with, each of the said eight thousand bonds, a certificate of the additional security of the said Land Grant Mortgage; which certificate shall be signed by the President of said party of the second part, the Vice-President of said Continental Improvement Company, and the said trustees, respectively, and shall be in the form following, to wit:

CERTIFICATE OF ADDITIONAL SECURITY OF LAND GRANT
MORTGAGE.

"*Know all men by these presents,* that in addition to being secured by a deed of trust or mortgage upon the Grand Rapids and Indiana Railroad, mentioned in the annexed bond, the payment of the said bond, and of the several coupons thereunto belonging, as well as of the other bonds of the series of which the annexed bond is one, and of the coupons thereunto belonging, as the same respectively become due, is secured by a certain deed of trust or mortgage, bearing date on the first day of October, 1869, duly executed and delivered by the Grand Rapids and Indiana Railroad Company, in the said bond mentioned, and the Continental Improvement Company, a corporation of the State of Pennsylvania, conveying, to John Edgar Thomson and George W. Cass, trustees, all the right, title and interest of the said Grand Rapids and Indiana Railroad Company, and of the said Continental Improvement Company, as the grantees or vendees thereof, now held, or hereafter to be acquired, in and to all and singular all those certain lands, situate in the State of Michigan, which were heretofore granted by the Congress of the United States to the State of Michigan, and by the said State to the said Grand Rapids and Indiana Railroad Company, to aid in the construction of the said railroad; which said lands are estimated by said grants to comprise in the aggregate one million one hundred and sixty thousand (1,160,000) acres, more or less. And it is provided in the said last mentioned deed of trust or mortgage, that the said lands may be sold, from time to time, in the manner and upon the terms therein specified, and this bond be received in payment therefor, at its par value, and that all net proceeds of any and all sales of the same shall be applied, in conformity with the provisions of the said deed, to the purchase and redemption of the said bonds and coupons, for the benefit of the Continental Improvement Company, to which purpose all of the said net proceeds are inviolably pledged."

2. That the said party of the second part shall and will, in the construction of the Grand Rapids and Indiana Railroad aforesaid, connect the main track or tracks thereof, or cause the same to be connected, with the main track or tracks of the Pittsburgh, Fort Wayne and Chicago Railway aforesaid, at or near the City of Fort Wayne, in the State of Indiana, to wit, at a point distant about one thousand feet, westerly, from the railroad bridge now built across the St. Mary's River, at or near, and to the west, of, the said city; and that the said party of the second part shall and will thereafter, forever, maintain and keep up such connection with said railway, at the point or place aforesaid, in such manner that any and all cars which are or may be run upon the said railway can, upon the arrival thereof at the said point of connection, be run in, upon, and over the said railroad, to any or all points thereon, whenever the party of the first part or its as-

signs shall desire the same to be so transported, in the interchange
of business between said parties; and the party of the second
part shall and will, at all proper times, furnish the necessary
motive power to haul said cars over said railroad, from and to said
point of connection; and to enable such transfer of cars to be
made, if desired, as aforesaid, the said party of the second part
will forthwith cause the tracks of the said railroad to be laid and
maintained, of a compromise gauge, suitable for such interchange of business, according to the now existing gauge of the
principal or main track of the said railway; and a reasonable
car service shall be allowed to said party of the first part, or its
assigns, for the cars so run over said railroad.

3. That the said party of the second part shall and will, so
far and to such extent as it lawfully may, at all times hereafter,
forever influence, direct and send, by and over the said Pittsburgh, Fort Wayne and Chicago Railway, any and all passengers
who, and any and all freight and traffic which, may be transported over the said railroad, or any part thereof, to Fort Wayne,
or to the junction of said roads at the point or place aforesaid,
and the ultimate destination of whom, or which, shall be the City
of Pittsburgh, or the City of Chicago, or any place, lying between
said cities on route of the said railway. When such passengers and
traffic are destined to any points off the line of said railway
which can or may be reached via any other railroad owned,
operated, leased or otherwise controlled by the Pennsyvania
Railroad Company, then such passengers and traffic shall be
ticketed and manifested, so that they may be carried and transported via such line of railroad to the point of ultimate destination.

4. That the said party of the second part shall and will, from
time to time, fix and regulate the times of arrival at, and depart-
ure from, Fort Wayne aforesaid, of any and all trains for the
conveyance of passengers over said railroad, in such manner and
according to such time tables, that the said passenger trains
shall at all times closely connect at the said city, or at the aforesaid point of connection between said roads, with the passenger
trains of the said party of the first part: *Provided*, that nothing
herein contained shall be held to compel the party of the second
part to run on the said railroad more trains than may be reason-

ably required for the business thereof; but this article shall be held to relate to the running of such passenger trains as the business of said railroad may call for, or as shall be run thereon.

ARTICLE THIRD.—And the parties hereto further mutually promise and agree, that all fares and charges for the transportation over the said railway and railroad, in connection with each other, of such passengers and freight, respectively, carried over each of said roads, or any part of each thereof, by the way of Fort Wayne, or the connecting point aforesaid at or near said city, shall, from time to time, be divided and paid over to the said parties hereto, *pro rata*, according to the number of miles or distance upon the said railroad and railway, respectively, which such passengers or freight shall have been so carried; and each of the said parties hereto shall be entitled to issue tickets to passengers and bills for freight, in such form and condition therein as may, from time to time, be required and approved by the party of the first part and its assigns, requiring to be transported, in part over the one road and in part over the other, by way of Fort Wayne, or of such connecting point at or near said city,—for which tickets and bills, and the proceeds thereof, the said parties shall be bound to account to each other, from time to time, on such *pro rata* basis as aforesaid.

ARTICLE FOURTH.—And the said party of the first part further promises and agrees, to and with the said party of the second part, that the said party of the first part will not in any unreasonable manner whatever discriminate against the said party of the second part, in favor of any other party, in the direction or control of any traffic between Fort Wayne, or the connecting point aforesaid, at or near said city, and any point or place on the line of, or usually accessible by, the said railroad.

ARTICLE FIFTH.—It is hereby expressly declared, that this contract is entered into by the party of the first part, in contemplation of the assignment of the interest of the said party of the first part, in and under the same, to the Pennsylvania Railroad Company aforesaid, and the assumption by the said last mentioned Company, of all obligations therein contained, on the part and behalf of the said party of the first part; and the party of the second part hereby assents to such assignment, and agrees that, thereupon, the said Pennsylvania Railroad Company shall be deemed to have assumed and taken the place of the party of the first part, under these presents, in the same manner and with the same effect as if the said Pennsylvania Railroad Company were the party of the first part hereto, instead of the said Pittsburgh, Fort Wayne and Chicago Railway Company; and all the provisions of these presents, which confer rights or impose obligations upon the party of the first part hereto, shall thereafter apply and be deemed to relate to the said assignee.

    IN WITNESS WHEREOF, the said parties of the first and second parts have caused their respective corporate seals to be hereunto affixed, and the same to be attested by the signatures of their respective Presidents and Secretaries, the day and year first above written.

The Pittsburgh, Fort Wayne and Chicago Railway Company,

By

President.

Secretary.

The Grand Rapids and Indiana Railroad Company,

By

President.

Secretary.

Signed, sealed and delivered, }
   in the presence of   }

ASSIGNMENT AND ASSUMPTION OF CONTRACT WITH THE GRAND RAPIDS AND INDIANA RAILROAD COMPANY.

THIS INDENTURE, made the first day of October, in the year one thousand eight hundred and sixty-nine, between the PITTS- BURGH, FORT WAYNE AND CHICAGO RAILWAY COMPANY, party of the first part, and the PENNSYLVANIA RAILROAD COMPANY, party of the second part:

Whereas, the said party of the first part, as contemplated by Article Thirteenth of the lease of the Pittsburgh, Fort Wayne and Chicago Railway, heretofore executed to the party of the second part, has concluded a certain contract with the Grand Rapids and Indiana Railroad Company, relative to the construction and operation of the Grand Rapids and Indiana Railroad, the connection thereof with the said Pittsburgh, Fort Wayne and Chicago Railway, and the negotiation and guaranty of certain First Mortgage Bonds, to be issued by the said Grand Rapids and Indiana Railroad Company, for the purpose of procuring the means of constructing the said railroad,—as by reference to said contract, which bears date on the thirtieth day of September, 1869, and a copy whereof is hereunto annexed, will more fully and at large appear.

And whereas, the said contract between the party of the first part and the said Grand Rapids and Indiana Railroad Company has been, in conformity to the lease aforesaid, duly submitted to, and the form thereof approved by, the Board of Directors of the said party of the second part hereto.

And whereas, a deed of trust or mortgage, bearing even date herewith, has been, in like conformity to said lease, duly executed by the said Grand Rapids and Indiana Railroad Company, jointly with the Continental improvement Company, a corporation of the State of Pennsylvania, to John Edgar Thomson and George W. Cass, trustees, conveying to the said trustees all the right, title, and interest of the said Grand Rapids and Indiana Railroad Company, as well as all the right, title, and interest of the said Continental improvement Company—the interest of the last-named Company being derived from the said Grand Rapids and Indiana

Railroad Company, under a contract for the construction of the. said railroad—in and to all and singular those certain lands, situate in the State of Michigan, which were heretofore granted by the Congress of the United States to the said State of Michigan, and by the said State of Michigan to the said Grand Rapids and Indiana Railroad Company, to aid in the construction of its railroad; such right, title, and interest in the said lands being conveyed, as aforesaid, to the said trustees, upon and subject to the trusts and agreements in the said deed of trust or mortgage mentioned and set forth, for the purpose of more effectually securing the payment of the bonds aforesaid, to be issued by the said Grand Rapids and Indiana Railroad Company, to an aggregate amount not exceeding eight millions of dollars.

And whereas, the said Grand Rapids and Indiana Railroad Company, in like conformity to said lease, has also transferred, or caused to be transferred and assigned, to the said John Edgar Thomson, Herman J. Lombaert and Edmund Smith, as trustees, a majority of the outstanding capital stock of the said Grand Rapids and Indiana Railroad Company, to be held by the said trustees as collateral security for the benefit and protection of the said party of the first part hereto, and its assigns, against any and all loss, damage and liability whatsoever, which can or may arise or happen to the said party of the first part, or its assigns, upon, under, on account, or by reason of the agreement in the said annexed contract set forth, for the purchase at par of the bonds aforesaid, numbered from 1 to 4,000, inclusively, and the coupons thereto annexed, if such bonds or coupons should remain due and unpaid for sixty days after the maturity thereof ; or upon, under, on account, or by reason of the memorandum or certificate of such agreement to the said four thousand bonds annexed—as by reference to a certain instrument, bearing date on the thirtieth day of September, 1869, executed by the said Grand Rapids and Indiana Railroad Company to the said trustees, will more fully and at large appear :

Now, THEREFORE, THIS INDENTURE WITNESSETH, that the said party of the first part, for and in consideration of the premises, and of the agreements, on the part and behalf of the said party of the second part, hereinafter set forth, to be kept and performed, has bargained and sold, and by these presents does grant, bargain, sell, assign, transfer and set over unto the said party of the second part, ALL THE RIGHT, TITLE, INTEREST, claim, property and benefit of the said party of the first part, in, to and under the said contract, bearing date on the thirtieth day of September, 1869, made by the said party of the first part with the said Grand Rapids and Indiana Railroad Company, and a copy whereof is hereto annexed, as aforesaid ; and also all the right, title, interest and benefit of the said party of the first part, in, to and under the said trust, issue or deposit of the said capital stock : *Provided always*, and upon this express condition, *that the said party of the second part shall and does assume and take the place of* the said party of the first part, under the said annexed contract, and indemnify the said party of the first part from and against any and all liability on account thereof.

And the said party of the second part, in consideration of the premises, and of the fulfillment—which is hereby acknowledged—of the conditions in Article Thirteenth of said lease mentioned, in reference to the assumption of said contract, as well as the sum of one dollar to it duly paid, has assumed and undertaken, and does hereby assume and undertake, the performance of the said contract with the said Grand Rapids and Indiana Railroad Company, on the part and behalf of the said party of the first part, according to the true intent and meaning thereof ; and has promised and agreed, and does hereby promise and agree, to and with the said party of the first part, for the benefit of the person or persons who may hereafter become the holder or holders of the several bonds and coupons of the Grand Rapids and Indiana Railroad Company, of the series of the First Mortgage Bonds thereof, numbered from 1 to 4,000, inclusively, that the said party of the second part shall and will purchase, pay for, and take, at the par value thereof, upon demand, from the holder or holders of the same, any and all of the bonds or coupons of the said series, which shall or may remain unpaid for sixty days, after the maturity and demand of payment thereof, and after the service upon the said party of the second part of the notice, in writing, provided in the said Article to be given to the said party of the first part, or the assigns, in such contingency ; *Provided, always*, that the bonds and coupons so purchased shall continue in full force and effect in the hands of the said party of the second part after such purchase, as in said Article specified. And the said party of the second part shall and will observe and do every matter and thing mentioned in the said contract, to be by the said party of the first part observed and done.

And the said party of the second part, for the consideration aforesaid, hereby further covenants, promises and agrees to and with the said party of the first part, that the said party of the second part shall, and will, at any and all times hereafter, save and keep harmless and indemnified the said party of the first part, of, from, and against any and all costs, damages, suits, actions and causes of action, expenses, liabilities, claims and demands whatsoever, which can or may arise, or be brought against, or incurred by the said party of the first part, under or on account of the said contract with the said Grand Rapids and Indiana Railroad Company, or anything therein contained, or under or on account of any matter or thing done, or to be done by the said party of the first part pursuant thereto.

> IN WITNESS WHEREOF, the said parties of the first and second parts have caused their respective corporate seals to be hereunto affixed, and the same to be attested by the signatures of their respective Presidents and Secretaries, the day and year first above written.

Signed, sealed and delivered }
  in the presence of   }

The Pittsburgh, Fort Wayne and Chicago Railway Company,

      by

                              President.

                              Secretary.

The Pennsylvania Rail Road Company,

      by

                              President.

                              Secretary.

PAPERS

RELATING TO THE GUARANTEED SPECIAL STOCK

ISSUED TO REPRESENT

PERMANENT IMPROVEMENTS,

AND BEARING DIVIDENDS

PAYABLE BY THE LESSEE.

[*Note.*—The Sixteenth Article of the Lease provides, that for the purpose of enabling the lessee to meet the obligations of the Company to the public, by making, from time to time, such improvements upon and additions to the Railway itself in the extension of facilities for increased business "*by additional tracks and depots, shops and equipments, and the substitution of stone or iron bridges for wooden bridges, or steel rails for iron rails*," the party of the first part shall issue to the lessee a *Special Stock ;* but the dividends on this special stock are to be payable *by the lessee*, and not by the Pittsburgh, Fort Wayne and Chicago Railway Company. Their payment is not to infringe, in any respect, upon the rights of the stockholders to whom the annual dividend fund of $1,380,000 is guaranteed. The *lessee* is to pay these dividends, "without deduction from the rent hereinbefore reserved." Not only this, but the stock itself can be issued only "in respect to improvements of and additions to the said railway, which, *and estimates and specifications* of which, shall have been submitted to and approved by the said party of the first part, in writing; *and all such improvements and additions shall be made in such manner as shall be approved by the said party of the first part*. (Vide Lease, Art. 16.)

The Board of Directors have exercised great care in respect to issues already made of this betterment stock. They have employed Mr. JOHN B. JERVIS, perhaps the most competent man in the United States for that purpose, to see that the improvements made were of the character represented, and such as called for the issue of this stock under the lease. His report being satisfactory, the outstanding issues were made. In making them, however, it will be seen, by the documents below, including the form of the special stock certificate and the copy of contract printed on the back, that the interests of the stockholders, for whose benefit the lease was made, have been surrounded with every safeguard.

In connection with this Special Improvement Stock, the fact should be borne in mind that, under the provisions of the Sinking Funds, the First and Second Mortgage Bonds will be extinguished in about twenty years, and that, under the terms of the certificate and agreement by which it is issued, the payment of dividends at the rate of seven per centum per annum, payable quarterly, is absolutely promised by the Pennsylvania Railroad Company.]

No.   [Vignette.]   Shares.

GUARANTEED SPECIAL STOCK.

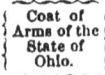

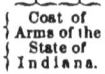

18

Registrar of Transfers.

day of

Countersigned and Registered this   The Third National Bank of the City of New York.

By

Issued under the Lease of the Pittsburgh, Fort Wayne and Chicago Railway, to the Pennsylvania Railroad Company, and entitled to quarterly dividends, free of taxes, at the rate of seven per cent. per annum, guaranteed by the Pennsylvania Railroad Company as Lessee.

This is to Certify, that the owner of   shares of One Hundred Dollars each, in the Special Improvement Stock issued by Pittsburgh, Fort Wayne and Chicago Railway Company in pursuance of the provisions of Article Sixteenth of the Lease of the Pittsburgh, Fort Wayne and Chicago Railway to the Pennsylvania Railway Company, which Lease bears date on the 7th day of June, 1869, and which Special Improvement Stock is by agreement with the Pennsylvania Railroad Company, dated October 28th, 1871, entitled to dividends at the rate of seven per cent. per annum, clear of all taxes, and the payment thereof guaranteed to the holder by the said Pennsylvania Railroad Company as Lessee, the dividends to be payable quarterly, namely : on the first days of January, April, July and October in each year, free of taxes, after duly providing for the payment of the regular quarterly dividends on the general or prior stock, including under the designation, "general or prior stock" all the stock, other than the said guaranteed special stock which the said Railway Company had issued or authorized to be issued before the date of the contract of which a copy is endorsed hereon, as appears by the provisions of said contract, dated October 28th, 1871, of which a copy is endorsed hereon. The special stock represented hereby is in all respects subject to the said general or prior stock, and the right of the holders of said general or prior stock to have distributed to them in quarterly instalments an annual dividend fund of $1,380,000 free of all taxes.

This stock is transferable only upon the books of said Pittsburgh, Fort Wayne and Chicago Railway Company by the said owner in person or by attorney, and upon the surrender of this certificate. This certificate shall not be valid until countersigned by the Transfer Agents of said Company, in the City of New York, and also by the Third National Bank, the Register of Transfers thereof in said city.

> IN WITNESS WHEREOF, the said Pittsburgh, Fort Wayne and Chicago Railway Company has caused its corporate seal to be hereunto affixed, and the same to be attested by the signatures of its President and Secretary this   day of   18

President.

Secretary.

Countersigned this   day }
· of   18   {

Transfer Agents.

[Endorsed :]

COPY OF AGREEMENT OF THE PENNSYLVANIA RAILROAD COMPANY.

The Pennsylvania Railroad Company having, in pursuance of Article Sixteenth of the Lease of the Pittsburgh, Fort Wayne and Chicago Railway, requested the preparation and issue to it from time to time, of a special stock to be designated the Guaranteed Special Stock of said railway company, the said Pennsylvania Railroad Company hereby promises, agrees and guarantees to and with the Pittsburgh, Fort Wayne and Chicago Railway Company, for the benefit of each and every person who may become a holder of said stock, after the same is issued to said Pennsylvania Railroad Company or to its order ; that the said Pennsylvania Railroad Company, its successors or assigns, shall and will provide and pay to the said Pittsburgh, Fort Wayne and Chicago Railway Company, quarterly, to wit: on or before the first days of January, April, July and October, in each and every year, an amount sufficient to pay quarterly dividends at the rate of seven per centum per annum, upon said special stock, as free of taxes as the dividends upon the general or prior stock of said Pittsburgh, Fort Wayne and Chicago Railway Company are made payable by the provisions of said lease, which payment shall be made quarterly in each year, after providing for, and in addition to fully paying or providing for the payment of the regular quarterly dividends payable upon the general or prior stock of the said railway company, including under the designation "general or prior stock" all the stock, other than the said special stock, which the said railway company has heretofore issued or authorized to be issued, for which general or prior stock a dividend fund of $1,380,000 per annum is inviolably pledged and set apart under the provisions of said lease, bearing date June 7, 1869. And it is further agreed, that all needful expenses connected with the said guaranteed special stock, the issue of certificates therefor, and the payment of dividends thereon, shall be borne and paid by the Pennsylvania Railroad Company.

The promises and agreements hereinbefore set forth are made by the said Pennsylvania Railroad Company, not only to and with the said Pittsburgh, Fort Wayne and Chicago Railway Company, but to and with each and every person who shall become a holder of the said special guaranteed stock.

A copy of this agreement shall be printed on the back of each certificate, and the Transfer Agents of the Pittsburgh, Fort Wayne and Chicago Railway Company may certify, and they are hereby authorized to certify on behalf of the Pennsylvania Railroad Company, as well as of the Pittsburgh, Fort Wayne and Chicago Railway Company, on each and every certificate for guaranteed special stock, issued from time to time under the provisions hereof, that this agreement has been duly executed by the Pennsylvania Railroad Company, under its corporate seal.

>IN WITNESS WHEREOF, the said Pennsylvania Railroad Company has caused its corporate seal to be hereunto affixed, and these presents to be signed by its President and Secretary, the 28th day of October, 1871.

>J. EDGAR THOMSON,
>*President.*

Jos. LESLEY,

Secretary.

```
Corporate Seal
     of
the Pennsylvania
  Railroad Co.
```

For, and in behalf of the Pennsylvania Railroad Company, as well as the Pittsburgh, Fort Wayne and Chicago Railway Company, hereby certify that the agreement, of which the foregoing is a true copy, has been duly executed by the Pennsylvania Railroad Company, with its corporate seal attached, under date October 28th, 1871.

>*Transfer Agent.*

Know all men by these presents, That , for value received, have bargained, sold, assigned and transferred, and by these presents do bargain, sell, assign, and transfer unto ,
shares of the Guaranteed Special Stock, issued by the Pittsburgh, Fort Wayne and Chicago Railway Company, standing in name, on the books of the said Company, and represented by the within certificate; and do hereby constitute and appoint . true and lawful attorney , irrevocable for and in name and stead, but to use, to sell, assign, transfer, and set over all or any part of the said stock, and for that purpose, to make and execute all necessary acts of assignment and transfer, and one or more persons to substitute with like full power, hereby ratifying and confirming all that

said attorney or   substitute or substitutes shall lawfully do by virtue hereof.

IN WITNESS WHEREOF,   have hereunto set   hand and seal, the   day of   18 .

Sealed and delivered in presence of

[SEAL.]

PITTSBURGH, FORT WAYNE AND CHICAGO RAILWAY COMPANY.

No.   Shares.

GUARANTEED SPECIAL STOCK,

Representing the actual cash value of property placed upon the line by the Lessees and issued under the Lease of the Pittsburgh, Fort Wayne and Chicago Railway to the Pennsylvania Railroad Company, and entitled to quarterly dividends, free of taxes, at the rate of seven per cent. per annum, guaranteed by the Pennsylvania Railroad Co. as Lessee.

THIS IS TO CERTIFY, that   the owner of   shares of one hundred dollars each in the Special Improvement Stock issued by the Pittsburgh, Fort Wayne and Chicago Railway to the Pennsylvania Railroad Company, which lease bears date on the 7th day of June, 1869, and which Special Improvement stock is by agreement with the Pennsylvania Railroad Company, dated October 28th, 1871, entitled to dividends at the rate of seven per cent. per annum, clear of all taxes, and the payment thereof guaranteed to the holder by the said Pennsylvania Railroad Company as Lessee, the dividends to be payable quarterly, namely, on the first days of January, April, July and October in each year, free of taxes after duly providing for the payment of the regular quarterly dividends on the general or prior stock, including under the designation "general or prior stock," all the stock other than the said Guaranteed Special Stock, which the said Railway Company had issued, or authorized to be issued, before the date of the contract, of which a copy is endorsed hereon, as appears by the agreement of said Company, dated October 28th, 1871, of which a copy is endorsed hereon. The Improvement Stock represented hereby is in all respects subject to the said general or prior stock, and the right of the holders of said general or prior stock to have distributed to them in quarterly instalments an annual dividend fund of $1,380,000, free of all taxes.

This stock is transferable only upon the books of said Pittsburgh, Fort Wayne and Chicago Railway Company, by the said owner, in person or by attorney, and upon the surrender of this certificate.

This certificate shall not be valid until countersigned by the Transfer Agents of said Company, in the City of New York, and also by the Third National Bank, the Register of Transfers thereof, in said city.

> In WITNESS WHEREOF, the said Pittsburgh, Fort Wayne and Chicago Railway Company has caused its corporate seal to be hereunto affixed, and the same to be attested by the signatures of its President and Secretary, this   day of   18

President.

Secretary.

Countersigned and registered this }
          day of   -   18 . }

THE THIRD NATIONAL BANK
          OF THE CITY OF NEW YORK,
By

Register of Transfers.

Countersigned this   day of }
               18   .   }

Transfer Agents.

[Endorsed.]

COPY OF AGREEMENT OF THE PENNSYLVANIA RAILROAD COMPANY.

The Pennsylvania Railroad Company having, in pursuance of Article Sixteenth of the Lease of the Pittsburgh, Fort Wayne and Chicago Railway, requested the preparation and issue to it from time to time of a special stock, to be designated the Guaranteed Special Stock of said Railway Company, the said Pennsylvania Railroad Company hereby promises, agrees and guarantees to and with the Pittsburgh, Fort Wayne and Chicago Railway Company, for the benefit of each and every person who may become a holder of said stock after the same is issued to said Pennsylvania Railroad Company, or to its order, that the said Pennsylvania Railroad Company, its successors or assigns, shall and will provide and pay to the said Pittsburgh, Fort Wayne and Chicago Railway Company, quarterly, to wit: on or before the first days of January, April, July and October, in each and every year, an amount sufficient to pay quarterly dividends, at the rate of seven per centum per annum, upon said special stock, as free of taxes as the dividends upon the general or prior stock of said Pittsburgh, Fort Wayne and Chicago Railway Company are made payable by the provisions of said lease, which payment shall be made quarterly in each year, after providing for, and in addition to fully paying or providing for the payment of the regular quarterly dividends, payable upon general or prior stock of the said Railway Company, including under the designation, "general or prior stock," all the stock, other than the said special stock, which the said Railway Company has heretofore issued or authorized to be issued ; for which general or prior stock a dividend fund of $1,380,000 per annum is inviolably pledged and set apart under the provisions of said lease, bearing date June 7, 1869. And it is further agreed, that all needful expenses connected with the said Guaranteed Special Stock, the issue of certificates therefor, and the payment of dividends thereon, shall be borne and paid by the Pennsylvania Railroad Company.

The promises and agreements hereinbefore set forth are made by the said Pennsylvania Railroad Company, not only to and with the said Pittsburgh, Fort Wayne and Chicago Railway

Company, but to and with each and every person who shall become a holder of the said Guaranteed Special Stock. A copy of this agreement shall be printed on the back of each certificate, and the Transfer Agents of the Pittsburgh, Fort Wayne and Chicago Railway Company may certify, and they are hereby authorized to certify, on behalf of the Pennsylvania Railroad Company, as well as of the Pittsburgh, Fort Wayne and Chicago Railway Company, on each and every certificate for Guaranteed Special Stock, issued from time to time under the provisions hereof, that this agreement has been duly executed by the Pennsylvania Railroad Company under its corporate seal.

> In witness whereof, the said Pennsylvania Railroad Company has caused its corporate seal to be hereunto affixed, and these presents to be signed by its President and Secretary, the 28th day of October, 1871.

J. EDGAR THOMSON,

President.

Jas. Lesley,

    Secretary.

```
.....................
: Corporate Seal :
:     of the      :
:  Pennsylvania   :
: Railroad Com'y. :
.....................
```

For and in behalf of the Pennsylvania Railroad Company, as well as the Pittsburgh, Fort Wayne and Chicago Railway Company,   hereby certify that the agreement of which the foregoing is a true copy, has been duly executed by the Pennsylvania Railroad Company, with its corporate seal attached, under date October 28th, 1871.

Transfer Agent.

Know all men by these presents, That

for value received have bargained, sold, assigned and transferred, and by these presents do bargain, sell, assign and transfer unto

shares of the Guaranteed Special Stock issued by the Pittsburgh,
Fort Wayne and Chicago Railway Company, standing in
name, on the books of the said Company, and represented by the
within certificate; and   do hereby constitute and appoint

true and lawful attorney, irrevocable, for   and in
name and stead, but to   use, to sell, assign, transfer and set
over all or any part of the said stock, and for that purpose to
make and execute all necessary acts of assignment and transfer,
and one or more persons to substitute with like full power, hereby ratifying and confirming all that   said attorney or
substitute or substitutes shall lawfully do by virtue hereof.

IN WITNESS WHEREOF   have hereunto set
hand and seal the   day of
18

Sealed and delivered }
in presence of   {

[SEAL.]

RENEWAL
OF
EQUIPMENT BONDS,
IN
CONFORMITY WITH THE LEASE.

UNITED STATES OF AMERICA.
STATES OF PENNSYLVANIA, OHIO, INDIANA AND ILLINOIS.

PITTSBURGH, FORT WAYNE AND CHICAGO RAIL-
WAY COMPANY.

No. $1,000.

EIGHT PER CENT. BOND,

Issued pursuant to the lease of the Pittsburgh, Fort Wayne and Chicago
Railway, in renewal of Equipment Bonds.

Know all men by these presents, That the Pittsburgh, Fort
Wayne and Chicago Railway Company is indebted to Charles
Lanier, of the City of New York, Trustee, or bearer, in the sum
of one thousand dollars, lawful money of the United States of
America, which sum the said Company promises to pay to the
said Lanier, Trustee, or to the bearer hereof, on the first day of
March, which will be in the year of our Lord one thousand eight
hundred and eighty-four, at the Office or Agency of the said
Company, in the City of New York, with interest at the rate of eight
per centum per annum, payable semi-annually, namely, on the first
days of March and September respectively in each year, until the
principal sum is paid, at the said Office or Agency, on the presentation and surrender of the annexed coupons as they severally
become due, free from any and all income taxes which may be
imposed by, or be payable to, the Government of the United
States on account thereof ; and it is hereby expressly declared
and agreed that in case default shall be made in the payment of
such interest, and the same shall remain unpaid and in arrear for
the space of three months, then this bond shall, at the option of
the owner thereof, become and be due and payable immediately.

This bond is one of a series of one thousand bonds of one
thousand dollars each, numbered consecutively, from 1 to 1,000,

inclusively, and all of similar tenor and date, which are issued by the said Pittsburgh, Fort Wayne and Chicago Railway Company, under and pursuant to the provisions of Article Second of the Lease of the Pittsburgh, Fort Wayne and Chicago Railway to the Pennsylvania Railroad Company, bearing date on the seventh day of June, A. D. 1869 ; and the payment of the interest on this bond as the same shall become due, and of the principal thereof at maturity, is assumed by the said PENNSYLVANIA RAILROAD COMPANY, under the terms and conditions of the said lease, and the series of bonds, of which this bond is one, is issued pursuant to the request, in writing, of the said Pennsylvania Railroad Company, made in conformity with the said lease, in order that these bonds may be substituted for a certain other issue of bonds, designated Equipment Bonds, which are mentioned and assumed to be paid in and by said lease, and for no other purpose.

This bond will pass by delivery, unless registered in the name of the owner, on the proper books of the said railroad company, either at New York or Pittsburgh; but this bond may be so registered upon said books, and upon such registration and the certification thereof hereon, by the proper officer or agent of said railway company, shall become transferrable only on the said books by the registered owner in person or by attorney.

This bond shall not become valid or obligatory until a certificate in authentication thereof, which is endorsed hereon, shall be duly signed by the said Trustee, or his successor in the trust.

IN WITNESS WHEREOF, the said railway company has caused its corporate seal to be hereunto affixed, and the same to be attested by the signatures of its

[L. S.] President and Secretary, and has also caused the coupons hereto annexed to be signed by its Secretary the first day of September, A. D. 1873.

President.

Secretary.

www.ingramcontent.com/pod-product-compliance
Lightning Source LLC
Chambersburg PA
CBHW021504210326
41599CB00012B/1138